The Physics of Thermoelectric Energy Conversion

The Physics of Thermoelectric Energy Conversion

H Julian Goldsmid
University of New South Wales (Emeritus)

Morgan & Claypool Publishers

Rights & Permissions
To obtain permission to re-use copyrighted material from Morgan & Claypool Publishers, please contact info@morganclaypool.com.

ISBN   978-1-6817-4641-8 (ebook)
ISBN   978-1-6817-4640-1 (print)
ISBN   978-1-6817-4643-2 (mobi)

DOI   10.1088/978-1-6817-4641-8

Version: 20170401

IOP Concise Physics
ISSN 2053-2571 (online)
ISSN 2054-7307 (print)

A Morgan & Claypool publication as part of IOP Concise Physics
Published by Morgan & Claypool Publishers, 40 Oak Drive, San Rafael, CA, 94903 USA

IOP Publishing, Temple Circus, Temple Way, Bristol BS1 6HG, UK

This book is dedicated to my wife, Joan.

Contents

Preface

For many years after their discovery in the early part of the nineteenth century the thermoelectric effects were not much more than a scientific curiosity. They eventually found application in the measurement of temperature and in the detection of thermal radiation. However, thermoelectricity is now regarded as a serious means for the conversion of heat into electricity and for heat pumping and refrigeration. The present-day thriving thermoelectric industry has come about through the dedicated research of materials scientists from all disciplines with a major contribution coming from solid state physicists. Considerable advances have been made since the introduction of semiconductor thermoelements in the middle of the twentieth century but much more needs to be done and it is hoped that this book will encourage a new generation of experimental and theoretical physicists to use their talents in the study of thermoelectricity.

Julian Goldsmid
Kingston Beach, Tasmania, Australia
March 2017

Acknowledgements

I acknowledge the support of Jeff Sharp and Jessica Bogart of II-VI Marlow in the preparation of the manuscript.

Author biography

H Julian Goldsmid

After graduating from Queen Mary College, University of London in 1949 Julian Goldsmid joined the scientific staff of the General Electric Company in their Wembley laboratories. In 1964 he was appointed Reader in Solid State Physics at the University of Bath and in 1969 he became Professor of Experimental Physics at the University of New South Wales in Sydney. After retiring from his chair in1988 he was appointed Emeritus Professor. From 1989 to 1995 he was Chairman of the Australian National Standards Commission.

Julian Goldsmid has been a Visiting Professor at the universities of Karlsruhe, Sussex, Southampton and the Southern Methodist University in Dallas. Until recently he was a consultant for Marlow Industries. He is a Fellow of the Institute of Physics and Honored Academician of the International Thermoelectric Academy. In 2002 he was awarded the Golden Prize of that organisation. In 2012 he received the Outstanding Achievement Award of the International Thermoelectric Society. He also received the Lightfoot Medal of the Institute of Refrigeration in 1959. He was awarded his PhD degree by the University of London in 1958 and his DSc in 1966. He has been author or joint author of about 200 publications. He is credited with many developments in the field of thermoelectricity, including the first demonstration in 1954 of practical refrigeration using the Peltier effect in bismuth telluride.

List of symbols

A	Cross-section area, mean atomic weight
a	Lattice constant
B	Magnetic field
C	Thermal capacity
c	Diameter of defect
c_V	Specific heat per unit volume
D^*	Detectivity
d	Width
E	Electric field, energy
E_F	Fermi energy
E_g	Energy gap
e	Electronic charge
F_n	Fermi–Dirac integral
f	Fermi distribution function
f_0	Equilibrium Fermi distribution function
g	Density of electron states
h	Planck's constant
$\hbar$	$h/2\pi$
I	Electric current
I_ϕ	Current for maximum COP
i	Electric current density
j	Heat flux density
K	Thermal conductance
K_c	Thermal conductance per unit area of end plates
K_s	Transport integral
k	Boltzmann's constant
L	Length, Lorenz number
l_e	Mean free path of charge carriers
l_t	Mean free path of phonons
M	$(1 + ZT_m)^{1/2}$
m	Mass of free electron
m^*	Density-of-states effective mass
m_I	Inertial mass
m_N	Density-of-states mass for a single valley
N	Nernst coefficient, total number of modes of vibration, number of unit cell per unit volume, number of couples in a module, number of stages in cascade
N_A	Avogadro's number
N_v	Number of valleys in an energy band
n	Subscript for electrons
n	Electron concentration, ratio of layer thicknesses in a synthetic transverse thermoelement
P	Ettingshausen coefficient, Poisson's ratio
p	Subscript for positive holes
$\boldsymbol{p}$	Phonon momentum
q	Rate of heat flow, phonon wave number
Q_1	Rate of heat flow from source
R	Electrical resistance, gas constant, responsivity

R_H	Hall coefficient
R_L	Load resistance
r	Scattering law parameter
S	Righi–Leduc coefficient
s	Compatibility factor
T	Temperature
T_1	Temperature of heat source
T_2	Temperature of heat sink
T_m	Mean temperature, melting point
ΔT	Temperature difference
ΔT^*	Temperature difference between sink and source
ΔT_{max}	Maximum temperature difference
t	Time
u	Velocity of carriers
V	Voltage, mean atomic volume
v	Speed of sound
w	Electrical power
x	$\hbar\omega/kT$
Z	Thermoelectric figure of merit for couple
Z_{NE}	Thermomagnetic or Nernst–Ettingshausen figure of merit
z	Figure of merit for single material
z_d	Phonon drag figure of merit
z_{trans}	Transverse figure of merit
α	Seebeck coefficient
α_d	Phonon drag Seebeck coefficient
α_T	Thermal expansion coefficient
β	Chasmar and Stratton's materials parameter, phase difference
Γ	Gamma function
γ	Grüneisen's parameter
ε	Emissivity
ε_m	Parameter in melting rule
η	Efficiency, reduced Fermi energy
η_g	Reduced energy gap
θ_D	Debye temperature
κ	Thermal diffusivity
λ	Thermal conductivity
λ_e	Electronic thermal conductivity
λ_I	Thermal conductivity of insulation
λ_L	Lattice conductivity
μ	Carrier mobility
ν	Frequency
ξ	Reduced energy
π	Peltier coefficient
π_d	Phonon drag Peltier coefficient
ρ	Electrical resistivity, density
σ	Electrical conductivity
τ	Thomson coefficient, relaxation time
τ_0	Scattering law constant
τ_d	Relaxation time for phonon drag
τ_e	Relaxation time for charge carriers
τ_{eff}	Effective relaxation time for charge carriers

xiii

τ_N	Relaxation time for normal processes
τ_R	Relaxation time for umklapp processes
ϕ	Coefficient of performance, angle of transverse thermoelement with normal to layers
ϕ_{max}	Maximum coefficient of performance
χ	Compressibility
ω	Angular frequency
ω_D	Debye angular frequency

Chapter 1

The Seebeck and Peltier effects

1.1 Definition of the thermoelectric coefficients

In this book we shall review the physical aspects of an energy conversion process that does not require any mechanical movement. The thermoelectric phenomena in those solids that conduct electricity are a measure of the energy transported by the charge carriers. These effects are thermodynamically reversible but they are invariably accompanied by irreversible effects associated with electrical resistance and thermal conduction. One of our aims is to show how the reversible processes can be maximised and the irreversible processes minimised.

The effect that bears his name was discovered in 1821 by Thomas Seebeck. It is manifest as an electromotive force (EMF) or voltage, V, which appears when the junction between two dissimilar conductors (A and B) is heated. Strictly speaking, the effect depends on the temperature difference, ΔT, between the two junctions that are needed to complete the electrical circuit. The Seebeck coefficient is defined as

$$a_{AB} = \frac{V}{\Delta T}. \tag{1.1}$$

Although the effect occurs only when there is a junction between two materials, the Seebeck effect is a characteristic of the bulk rather than surface properties.

The Seebeck effect is often used in the measurement of temperature. A thermocouple for this purpose typically consists of two metals or metallic alloys. For example, copper and constantan, with a differential Seebeck coefficient of about $40~\mu V~K^{-1}$, are commonly employed. Since a thermocouple measures the temperature difference between two junctions, the second junction must be placed in an enclosure at some known temperature, such as a bath of melting ice at 0 °C.

doi:10.1088/978-1-6817-4641-8ch1

A closely related effect was discovered in 1834 by Jean Peltier. Peltier heating or cooling occurs when an electric current, I, flows through the junction between two conductors. The Peltier coefficient, π_{AB}, is defined as

$$\pi_{AB} = \frac{q}{I}, \tag{1.2}$$

where q is the rate of heating or cooling.

It is rather more difficult to demonstrate the Peltier effect than the Seebeck effect. If the thermocouple branches are metallic, the reversible Peltier effect is usually overshadowed by irreversible Joule heating. Thus, unless the electric current is very small, the best that can be done is to show that the overall heating is less for the current flow in one direction rather than the other. Of course, with the semi-conductor thermoelements that are now available, it is easy to show that water can be frozen with the current in one direction and boiled with the current in the opposite direction.

1.2 The Kelvin relations

It is not surprising that the Seebeck and Peltier coefficients are inter-dependent. Kelvin established two laws that relate α_{AB} and π_{AB} to one another and to a third quantity, the Thomson coefficient, τ. The Thomson coefficient is the rate of heating per unit length when unit current passes along a conductor for unit temperature gradient. It may be expressed, therefore, as

$$\tau = \frac{dq/dx}{IdT/dx}. \tag{1.3}$$

Unlike the Peltier and Seebeck effects, the Thomson effect exists for a single conductor and is present for both branches of a thermocouple.

The relationships between the thermoelectric coefficients can be determined by the principles of irreversible thermodynamics. These relationships, which are known as Kelvin's laws, are

$$\pi_{AB} = a_{AB}T, \tag{1.4}$$

and

$$\tau_A - \tau_B = T\frac{d\alpha_{AB}}{dT}. \tag{1.5}$$

It would clearly be helpful if we could assign Seebeck and Peltier coefficients to each branch of a thermocouple so that α_{AB} and π_{AB} would be equal to $(\alpha_A - \alpha_B)$ and $(\pi_A - \pi_B)$ respectively. We note that the thermoelectric coefficients are equal to zero for all pairs of superconductors, so it is reasonable to suppose that the absolute values of α and π are zero for any superconductor. This being so, we can obtain the absolute values of α and π for any normal conductor by joining it to a superconductor. Of course, this procedure is effective only below the critical temperature of the superconductor. However, the absolute Seebeck coefficient of the normal conductor can be extrapolated to higher temperatures [1, 2] using the second Kelvin relation, equation (1.5). This has actually been done for the metal lead, which can be used as a reference material in establishing the absolute Seebeck coefficients of other conductors.

The first of Kelvin's laws, equation (1.4), is useful in that it tells us that one does not need to specify both the Seebeck and Peltier coefficients. The Peltier coefficient is, in fact, rather difficult to determine, whereas the Seebeck coefficient is one of the easiest of physical properties to measure. Thus, it is common practice to develop the theory of thermoelectric energy conversion in terms of the Seebeck coefficient. If and when the Peltier coefficient is needed, it is replaced by αT.

1.3 Electrical resistance and thermal conductance

The Seebeck and Peltier effects are reversible and a thermoelectric energy convertor would have the characteristics of an ideal heat engine were it not for the presence of the irreversible effects of electrical resistance and heat conduction. If one attempts to reduce the Joule heating in a thermocouple by increasing the cross-section area and reducing the length one merely increases the heat conduction losses. In developing the theory of thermoelectric generation and refrigeration one needs to include terms that involve the electrical and thermal conductivities of the thermocouple materials.

The electrical conductivity, σ, is defined by the relation

$$\sigma = \frac{IL}{VA}, \tag{1.6}$$

where I is the electric current, V is the applied voltage, L is the length and A is the cross-section area. As we shall see later, it is important to specify that the voltage is that for isothermal conditions.

The other quantity of interest, the thermal conductivity, λ, is defined by

$$\lambda = \frac{qL}{A\Delta T}, \tag{1.7}$$

where q is the heat that is transported for a temperature difference ΔT over a length L.

It is obvious that the Seebeck and Peltier coefficients should be as high as possible and that the electrical conductivity should be large and the thermal conductivity small. However, as we shall see later, these requirements cannot all be met in a given material. Usually an increase in the thermoelectric coefficients is accompanied by a decrease in the electrical conductivity. In the next chapter we shall show the relative importance of the different properties and we shall introduce a quantity known as the figure of merit, z. We shall show how the performance of thermoelectric refrigerators and generators is related to z.

References

[1] Borelius G, Keesom W H, Johansson C H and Linde J O 1932 *Proc. Ned. Akad. Wet.* **35** 10
[2] Christian J W, Jan J P, Pearson W B and Templeton I M 1958 *Can. J. Phys.* **36** 627

Chapter 2

The thermoelectric figure of merit

2.1 Coefficient of performance of thermoelectric heat pumps and refrigerators

The possibility of using the Peltier effect as a means of refrigeration was recognised in the nineteenth century. A refrigerator is a device for transporting heat from a source at a temperature T_1 to a sink at a higher temperature T_2. The ratio of the cooling power to the rate of working, i.e. the electrical power input, is known as the coefficient of performance, ϕ. The value of ϕ generally falls as $(T_2 - T_1)$ increases and there is a maximum temperature difference, ΔT_{max}, that can be achieved with any particular heat engine. It is desirable that both ϕ and ΔT_{max} should be large. It was shown by Altenkirch in 1911 [1] that these quantities can be related to the Seebeck coefficient and to the ratio of the electrical to thermal conductivity in each of the branches of a thermocouple that is used as a refrigerator. Altenkirch's theory is still useful though, at the time that it was presented, no thermoelectric materials were available that allowed worthwhile values of ΔT_{max} to be reached.

We discuss the performance of a thermoelectric refrigerator using the simple model shown in figure 2.1. The diagram shows a single thermocouple situated between a source of heat and a heat sink. The couple will generally consist of a positive and a negative branch though occasionally one of the branches may be a metal with a Seebeck coefficient that is close to zero. For example, a super-conducting branch [2, 3] will not contribute to the thermoelectric effects but neither will it impair the performance of the other, active, branch. The branches are usually linked by metallic conductors.

When an electric current I flows through the couple there will be a Peltier cooling effect $I\pi_{AB}$, which will be of the order of a tenth of a watt per amp. Most practical refrigerators will require several watts of cooling and this is achieved by connecting many thermocouples thermally in parallel and electrically in series so as to avoid the use of excessively large currents. However, the theory for a single couple is equally valid for a multi-couple arrangement.

doi:10.1088/978-1-6817-4641-8ch2

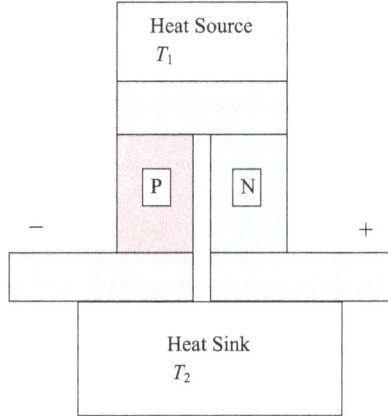

Figure 2.1. Basic thermoelectric refrigerator.

We shall suppose that the thermocouple makes perfect thermal contact with the source and sink and we shall ignore heat transport other than through the two branches. It will be necessary to take account of the relative ratios of length, L, to cross-section area, A, for the two arms. It will be supposed that the temperature dependence of the Seebeck coefficient and of the electrical and thermal conductivities can be ignored. This is a reasonable approximation for thermoelectric refrigerators since the temperature difference $(T_2 - T_1)$ is generally much less than the absolute temperature.

The Peltier cooling effect will be opposed by thermal conduction and by Joule heating, half of which will be delivered to the heat source and half to the sink. Thus, the overall rate of cooling is given by

$$q_1 = (\alpha_p - \alpha_n)T_1 I - I^2 R/2 - K(T_2 - T_1), \tag{2.1}$$

where R is the electrical resistance of the two branches in series and K is the thermal conductance of the branches in parallel. R and K are given by

$$R = \frac{L_p}{o_p A_p} + \frac{L_n}{o_n A_n}, \tag{2.2}$$

and

$$K = \frac{\lambda_p A_p}{L_p} + \frac{\lambda_n A_n}{L_n}. \tag{2.3}$$

The rate of expenditure of electrical energy, w, can be expressed as

$$w = (\alpha_p - \alpha_n)(T_2 - T_1)I + I^2 R. \tag{2.4}$$

Thence the coefficient of performance, ϕ, is given by

$$\phi = \frac{q_1}{w} = \frac{(\alpha_p - \alpha_n)T_1 I - I^2 R/2 - K(T_2 - T_1)}{(\alpha_p - \alpha_n)(T_2 - T_1)I + I^2 R}. \tag{2.5}$$

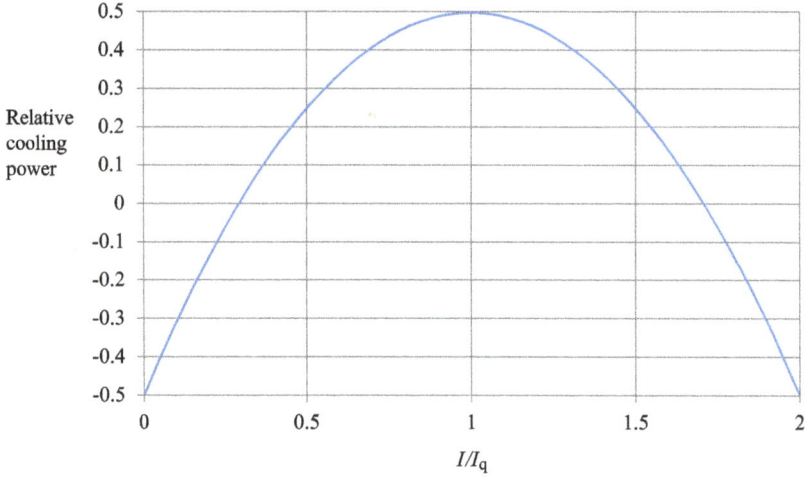

Figure 2.2. Plot of cooling power against electric current for a thermoelectric refrigerator. The current is given as a fraction of that which yields maximum cooling. The cooling power is expressed as a fraction of the maximum value if there were no temperature difference between the junctions. The actual temperature difference is arbitrarily chosen as $\Delta T_{max}/2$.

It will be noticed from equation (2.1) that the cooling power contains a term that increases linearly with the electric current and an opposing term that varies with the square of the current. Thus, there is a particular current at which the cooling power reaches a maximum. Also, because of heat conduction, the cooling power will be negative when the electrical current is very small. This is illustrated in the schematic plot of figure 2.2 in which the cooling power is plotted against the electric current.

2.2 The dimensionless figure of merit, ZT

It is evident from equation (2.5) that the coefficient of performance depends on the electric current. It reaches its maximum value [4] when the current is equal to I_ϕ where

$$I_\phi = \frac{(\alpha_p - \alpha_n)(T_2 - T_1)}{R\{(1 + ZT_m)^{1/2} - 1\}}. \qquad (2.6)$$

Here T_m is the mean temperature and Z is equal to $(\alpha_p - \alpha_n)^2/KR$. When the current is I_ϕ the coefficient of performance becomes

$$\phi_{max} = \frac{T_1\{(1 + ZT_m)^{1/2} - (T_2/T_1)\}}{(T_2 - T_1)\{(1 + ZT_m)^{1/2} + 1\}}. \qquad (2.7)$$

It is clear from this equation that the coefficient of performance rises as Z becomes larger and this quantity is known as the figure of merit. Nowadays it is more common to use the dimensionless figure of merit, ZT, rather than Z as a measure of the quality of a thermocouple.

For a given thermocouple, the figure of merit can be optimised by adjustment of the length and cross-section area of the branches. ZT reaches its largest value when KR has its minimum value. This occurs when

$$\frac{L_n A_p}{L_p A_n} = \left(\frac{\sigma_n \lambda_n}{\sigma_p \lambda_p}\right)^{1/2}. \tag{2.8}$$

When this condition is satisfied the figure of merit is given by

$$Z = \frac{(\alpha_p - \alpha_n)^2}{\{(\lambda_p/\sigma_p)^{1/2} + (\lambda_n/\sigma_n)^{1/2}\}^2}. \tag{2.9}$$

There is a useful relationship between the maximum temperature difference, ΔT_{max}, and the figure of merit. This relationship is

$$\Delta T_{max} = \frac{ZT_1^2}{2}. \tag{2.10}$$

This follows from equation (2.7) when we set the coefficient of performance equal to zero. The plot of ΔT_{max} against ZT_1, for a sink temperature of 300 K, is shown in figure 2.3. It will be seen that substantial temperature differences can be achieved when ZT_1 becomes of the order of unity.

It will be noted from its definition in equation (2.9) that the figure of merit involves a combination of the properties of the two branches. However, in searching for improved thermoelectric materials, it is convenient to consider the properties of each branch separately. Thus, it is usual to define a figure of merit, z, for an individual material, as $\alpha^2 \sigma/\lambda$. Since this figure of merit involves the square of the Seebeck coefficient, it is positive irrespective of whether the material is positive or negative. There is no simple relationship between the figure of merit, Z, for a couple

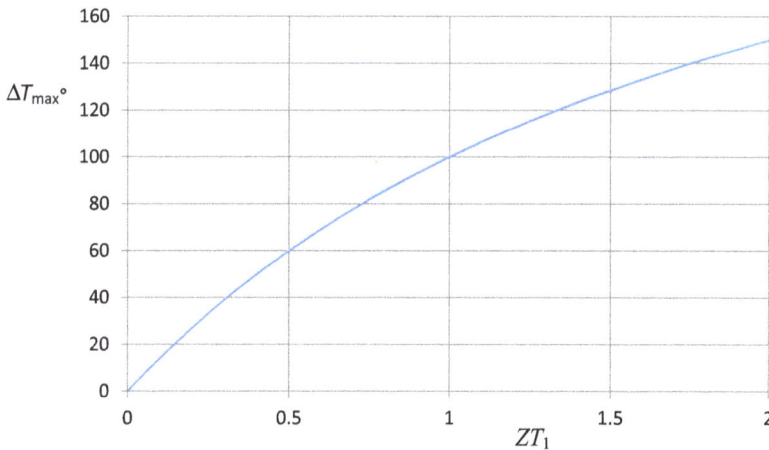

Figure 2.3. Plot of ΔT_{max} against ZT_1 for a thermoelectric refrigerator. The sink temperature, T_2, is equal to 300 K.

and the figures of merit z_p and z_n for the separate branches. However, it is often a good approximation to set Z equal to the mean of the values of z_p and z_n. Therefore, in discussing the selection and improvement of thermoelements, we shall invariably make use of the single-material figure of merit.

Although the Peltier effect finds its most common use in refrigeration, it should not be forgotten that a thermocouple can also be an effective heat pump. When a Peltier device is used in this mode, the temperature of the heat source may be greater than that of the sink. We can again make use of equation (2.1) with appropriate signs for each of the terms on the right hand side. Here the coefficient of performance is equal to the ratio of the rate of heating at source to the electrical power. The coefficient of performance can be considerably greater than unity, as it may also be for operation in the refrigeration mode.

2.3 The efficiency of thermoelectric generators

We turn now to the use of the Seebeck effect in the generation of electricity. We shall make the same assumption that there is no heat transfer between the source and sink other than through the thermocouple. Again, we expect that an actual generator will need a number of couples connected in series electrically and thermally in parallel. The output power would be too small for most applications if only one or two couples were employed. It will be supposed that the output from the generator is fed into a resistive load, R_L, the value of which can be adjusted. The arrangement is shown in figure 2.4.

We are interested in the efficiency, η, defined as the ratio of the power delivered to the load to the rate at which heat passes from the source to the sink. It must be remembered that the current produced by the Seebeck effect will itself lead to heat transfer between the source and sink through the Peltier effect. The electric current is given by

$$I = \frac{(\alpha_p - \alpha_n)(T_1 - T_2)}{R_L + R}. \tag{2.11}$$

The rate at which heat flows from the source is

$$q_1 = (\alpha_p - \alpha_n)IT_1 + K(T_1 - T_2) - \frac{1}{2}I^2R. \tag{2.12}$$

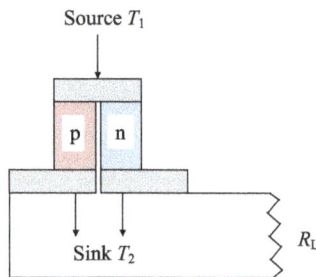

Figure 2.4. Simple thermoelectric generator connected to a resistive load, R_L.

and the power delivered to the load is

$$w = \left\{ \frac{(\alpha_p - \alpha_n)(T_1 - T_2)}{R_L + R} \right\}^2 R_L. \tag{2.13}$$

The efficiency is obtained as w/q_1 from equations (2.12) and (2.13).

The power output has its maximum value when the load resistance is equal to the internal resistance of the generator. However, if this condition is applied the efficiency can never exceed 50%. The efficiency is at its maximum when the load resistance is chosen so that

$$M = \frac{R_L}{R} = (1 + ZT_m)^{1/2}, \tag{2.14}$$

where T_m is the mean temperature. When R_L satisfies this equation, the efficiency has the value

$$\eta = \frac{(T_1 - T_2)(M - 1)}{T_1(M + T_2/T_1)}. \tag{2.15}$$

It will be noticed that, as M approaches infinity, the efficiency approaches the Carnot cycle value of $(T_1 - T_2)/T_1$. In figure 2.5 we show how the maximum efficiency varies with ZT_m for a heat sink temperature, T_2, of 300 K. It will be seen that, for ZT_m equal to unity and a temperature difference of 100 degrees, the efficiency is about 5%. This may be compared with an efficiency of 25% for a Carnot cycle working between the same temperatures.

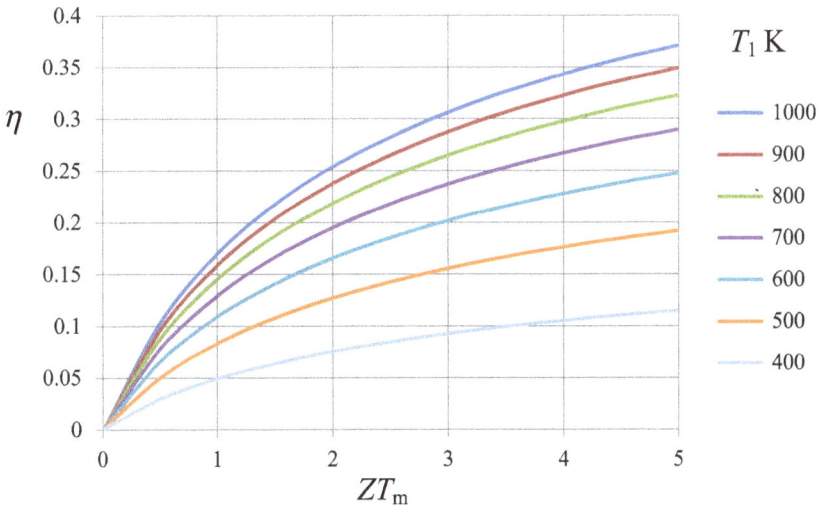

Figure 2.5. Plot of efficiency against ZT_m for thermoelectric generation. The heat sink temperature, T_2, is set at 300 K.

2.4 Multi-stage arrangements

If a thermoelectric generator is operating between widely different source and sink temperatures it is unlikely that a single pair of thermoelectric materials will cover the whole range. Thus, the generator may consist of a number of stages arranged in a thermal series. Alternatively, the branches of a single couple may be segmented, though this raises problems of compatibility.

Multi-stage units are sometimes employed in thermoelectric refrigeration if the required temperature difference is greater than can be achieved with a single stage. In principle, any required source temperature can be reached by this means but, in practice, it is difficult to obtain a temperature difference that is greater than, say, twice that for a single stage device.

Let us suppose that we have N stages. There is no reason why the coefficient of performance should be the same for all the stages but we shall assume that it is arranged for this to be the case and that for each stage the coefficient of performance is ϕ_1. Then the Nth and final stage will extract heat from the source at the rate q_1 and deliver heat at the rate $q_1(1 + 1/\phi_1)$ to the $(N-1)$th stage. This stage has to remove not only the heat from the source but also the power consumed by the Nth stage. Thus, the $(N-2)$th stage has to extract heat at the rate $q_1(1 - \phi_1)^2$. It is then found

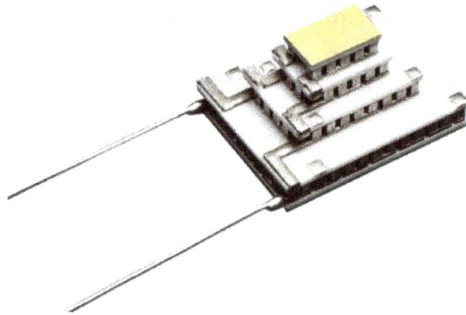

(a)

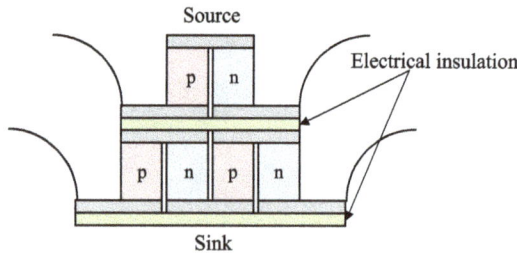

(b)

Figure 2.6. (a) Commercial four-stage cascade (courtesy of II-VI Marlow). (b) Schematic representation of a two-stage cascade. Note that the performance will fall short of the theoretical predictions because of thermal resistance between stages and heat loss through the electrical connections.

Table 2.1. ΔT_{max} for commercial multi-stage modules (data supplied by J Sharp of II-VI Marlow).

Number of stages	ΔT_{max} for zero load with $T_2 = 300$ K
1	73°
2	107°
3	123°
4	130°

that the rate at which heat is rejected by the first stage to the heat sink is $q_1(1 + 1/\phi_1)^N$. This must be equal to q_1 plus the total electrical power consumed. The overall coefficient of performance is then

$$\phi = \frac{1}{\left(1 + \frac{1}{\phi_1}\right)^N - 1}. \tag{2.16}$$

We illustrate the use of this relationship for a multi-stage cooler, or cascade, made up of thermocouples with ZT_m equal to 0.5 at all temperatures. This is a conservative value at room temperature but may well represent typical behaviour at the low-temperature end of the cascade.

Figure 2.6(a) portrays a commercial four-stage unit while figure 2.6(b) shows the essentials of a two-stage cascade. With the assumed value of ZT_m and T_2 equal to 300 K, the maximum temperature difference for a single stage would be about 55°. For such a temperature difference, the coefficient of performance would, of course, be zero. On the other hand, for a two-stage cascade with the same temperature difference, the value of ϕ_1 for each stage would be about 0.5 and the overall coefficient of performance would have the respectable value of 0.125. It is clear that there are advantages in using a cascade when the required temperature difference is close to the maximum for a single stage. Multi-stage Peltier coolers that yield temperatures well below 200 K are commercially available as shown in table 2.1.

References

[1] Altenkirch E 1911 *Phys. Z.* **12** 920
[2] Goldsmid H J, Gopinathan K K, Taylor K N R and Baird C A 1988 *J. Phys. D: Appl. Phys.* **21** 344
[3] Vedernikov M V and Kuznetsov V L 1994 *CRC Handbook of Thermoelectrics* ed D M Rowe (Boca Raton: CRC Press) p 609
[4] Ioffe A F 1957 *Semiconductor Thermoelements and Thermoelectric Cooling* (London: Infosearch) p 75

Chapter 3

Measuring the thermoelectric properties

3.1 Adiabatic and isothermal electrical conductivity

The measurement of the electrical conductivity of a thermoelectric material is not without its problems. If one measures the resistance between the ends of a sample, there is the danger of including a contribution from the contacts. This difficulty can be overcome if inset probes are used as in figure 3.1. However, it is not always necessary to use inset probes since it has been found that negligible contact resistances can be achieved for many thermoelectric materials provided that the sample length is not too small.

A more subtle problem is peculiar to thermoelectric materials [1]. The passage of an electric current not only produces a voltage drop due to the electrical resistance but it also leads to a temperature gradient through the Peltier effect. This temperature difference, in turn, produces a Seebeck voltage. This problem is most evident for good thermoelectric materials and, as we shall see later, can be used to advantage in the determination of the figure of merit. Here, however, we shall discuss the determination of the isothermal electrical conductivity.

The resistive voltage appears immediately on the introduction of an electric current but the Seebeck voltage takes some time to develop, depending on the thermal capacity of the system. Thus, if the electrical resistance is determined using an alternating current of sufficiently high frequency, it is the isothermal value that is obtained. Figure 3.2 shows an alternating current bridge with inset probes which has been used successfully on thermoelectric materials [2]. This apparatus was adapted from the classical Wheatstone bridge, using a vibration galvanometer rather than a direct-current galvanometer. The potential difference between the inset probes on the sample is compared with that between two points on the slide wire. Balance is obtained first with the switches in the S position and then with the switches moved to M. The variable resistors, B_1 and B_2, allow the balance points for the two probes to be located within the length of the slide wire. P and Q are standard resistors and the resistance per unit length of the slide wire is also a known quantity.

doi:10.1088/978-1-6817-4641-8ch3
3-1

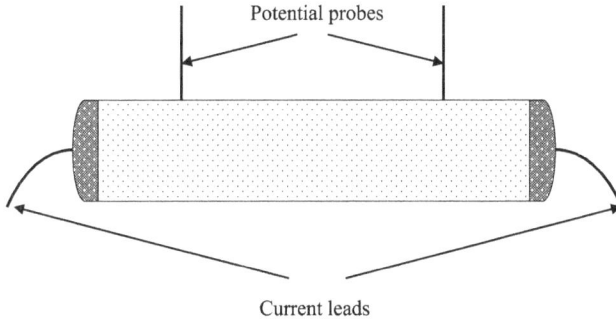

Figure 3.1. Use of inset probes in measuring the electrical conductivity.

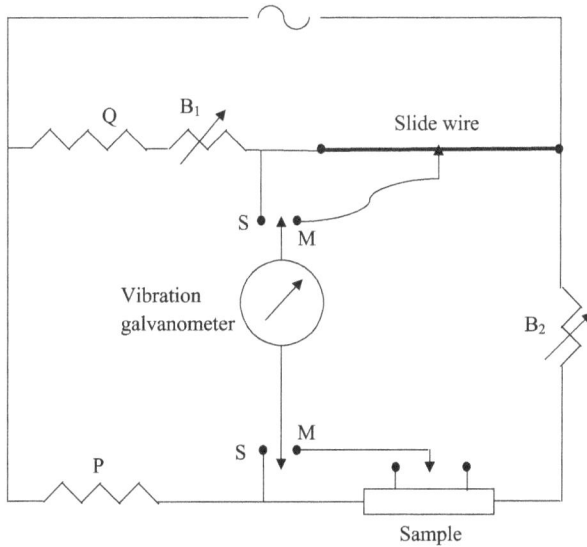

Figure 3.2. Alternating current bridge for measuring the electrical conductivity.

It is generally accepted that direct current measurements are more precise than those in which alternating current is employed. In particular, it is desirable to be able to use a direct current potentiometer if the Hall effect and the magnetoresistance are also being studied. Precise measurement of these effects has been carried out using a chopper system [3]. Here the current and the potential difference between probes attached to the sample are periodically reversed. The experimental arrangement is shown in figure 3.3. Additional probes are attached to the sample if the Hall coefficient is being measured.

One of the difficulties in using a chopping method is that transient voltages are generated upon switching the current. This difficulty is overcome by ensuring that the potential probes are isolated during the brief time that the switching actually occurs.

It is noteworthy that the four-probe method, which has been such a useful tool in the study of semiconductors, is also of some value for thermoelectric materials,

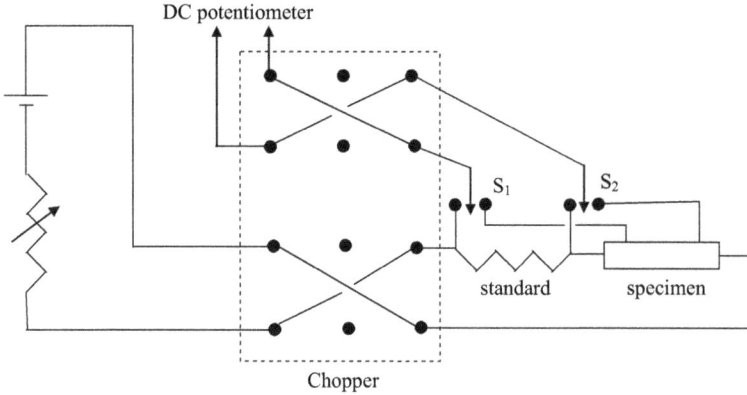

Figure 3.3. Chopper system for determining the isothermal resistivity. S_1 and S_2 are two-pole, two-way selector switches.

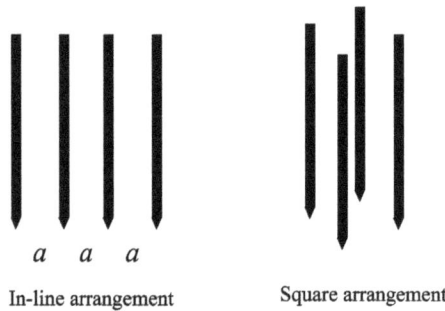

Figure 3.4. The four-probe method for determining the electrical conductivity.

though probably not when high precision measurements are needed. There are two arrangements as shown in figure 3.4. When the probes are in line the electrical conductivity is found from the relationship

$$\sigma = \frac{I}{2\pi a V},\tag{3.1}$$

where I is the current between the outer probes and a is the spacing between adjacent probes. V is the potential difference between the inner probes. Ideally, the measurement should be made with an alternating current.

Figure 3.4 also shows an alternative arrangement in which the probes are located at the corners of a square. The current is passed between probes on adjacent corners. This arrangement has the advantage that the Hall coefficient can also be determined by passing the current between diagonally opposite corners. The square configuration also allows somewhat smaller samples to be used, but the general rule applies that the specimen should be substantially greater than the space occupied by the probes.

One of the most important objections to the use of the four-probe method is that equation (3.1) is not applicable for anisotropic materials. A correction can be made if the anisotropic material is in the form of a single crystal but the method is of little use if it is polycrystalline. Unfortunately, some of the most widely used thermo-electric materials are in the form of anisotropic polycrystals.

3.2 Problems of measuring the thermal conductivity

It has long been recognised that it is much more difficult to make precise measure-ment of the thermal conductivity than the electrical conductivity. This is because there is no equivalent of an electrical insulator when we are dealing with heat conduction. Although it is true that there is no conduction of heat in a vacuum there is always heat transfer by radiation to be taken into account.

It may take a long time to reach thermal equilibrium in some systems but, even so, static methods have often been preferred over dynamic techniques for measuring the thermal conductivity, in the belief that they are more reliable. However, as we shall see, some dynamic methods are now regarded with favour.

In a steady-state method the sample is held between a heat source and a sink. The rate of heat flow can be determined from the power fed into the source or from the heat passing into the sink. These two alternatives will only be equivalent to one another if due account is taken of heat transfer by radiation and by convection and conduction through the surrounding air, if the space is not evacuated. It is necessary to determine the temperature gradient along the test sample. This can be done by finding the temperatures of the source and sink. Alternatively, the temperature at different points on the sample can be determined but it must be remembered that a thermometer pressed against the surface will not generally yield the true temper-ature. A more reliable result is obtained if the thermometer, usually a thermocouple, is embedded in a small hole, with contact being improved, perhaps, by spark welding.

There is also the question of the form factor for the sample. For electrical measurements, it is usually best to work with samples that have a large ratio of length to cross-section area. On the other hand, when the thermal conductivity is low, as it often is for good thermoelectric materials, a much smaller ratio of length to cross-section area is preferred. This sometimes means that different samples are needed for assessing the thermal and electrical conductivities. If different specimens are used, it is important that both are characteristic of the same material.

There are some advantages in using a comparison method rather than an absolute determination of the thermal conductivity. However, the comparison material should have a similar thermal conductivity to the material that is being tested and there is not always a standard substance for which the thermal conductivity is known with the required accuracy. Furthermore, comparison methods are usually slower than absolute techniques because of the increased length of the thermal path.

An example of lateral thinking being brought to bear on thermal conductivity measurement is to be found in the method used by Ioffe [4], one of the pioneers in the application of thermoelectricity. A discussion of Ioffe's system highlights some of the problems of thermal conductivity measurement and how to solve them. The principles of the method are shown in figure 3.5.

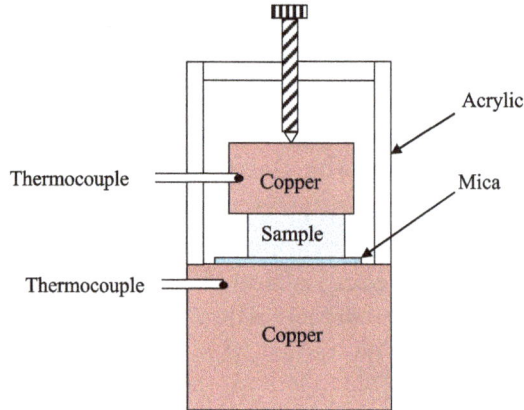

Figure 3.5. The Ioffe method for the measurement of thermal conductivity.

In the Ioffe method, the test specimen is sandwiched between two copper blocks. The acrylic shield attached to the larger block carries a screw that is used to ensure good contact on both faces of the sample. Electrical conduction between the sample and the lower block is prevented by a thin layer of mica but, of course, this introduces some undesirable thermal resistance. In use, the lower block is immersed in a cooling bath at some temperature, T_1. The temperature, T_2 of the upper block is then observed as a function of time. An approximate value for the thermal conductivity, λ, is found from the equation

$$-C_2\frac{\mathrm{d}T_2}{\mathrm{d}t} = \lambda(T_2 - T_1)\frac{A}{L}, \tag{3.2}$$

where C_2 is the thermal capacity of the upper block and A/L is the ratio of cross-section area to length for the sample. This equation does not take account of the fact that part of the heat entering the lower block comes from the sample. This factor can be included by adding one-third of the heat capacity of the sample to C_2. This correction requires knowledge of the specific heat of the substance under test, but this quantity does not have to be very accurate if the sample is much smaller than the upper block.

When equilibrium is established, it is found that the two copper blocks are still at different temperatures and this allows us to estimate the heat loss to the surroundings. There is also some transfer of heat through the air surrounding the sample and this can be determined by substituting a specimen of known thermal conductance for the test material. One must also correct for the thermal resistance between the sample and the two copper blocks, the correction being found by using different lengths, L. It has also been shown [5, 6] by a more refined theoretical treatment that one should neither make measurements too soon after immersing the lower block in the coolant nor when the upper block has come close to its equilibrium temperature.

In the search for new materials it is often necessary to find the thermal conductivity of very small specimens. Nowadays, this problem is solved by using the 3ω method [7]. In this technique, a thin metal strip is deposited on the sample as

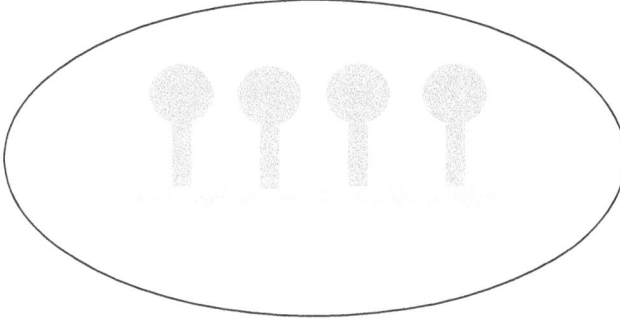

Figure 3.6. Experimental arrangement in the 3ω method. An alternating current at a frequency ω is introduced between the outer arms. The potential difference between the inner probes is found using a lock-in amplifier.

shown in figure 3.6. The strip is in good thermal contact with the substrate but electrically isolated from it. The outer arms are used for the introduction of an alternating electric current at a frequency ω while the inner arms pick up a potential difference. It is noted that the temperature dependence of the electrical conductivity allows the metal strip to act as a thermometer.

The electric current heats the strip, producing a temperature wave of frequency 2ω since heating occurs twice during each cycle. As the temperature wave travels radially outwards it suffers exponential damping. If the width of the strip is less than one-fifth of the thickness of the substrate, there is no significant interference from the reflected thermal wave. The observed potential difference will have a component with a frequency ω associated with the flow of current, and another component of frequency 2ω associated with the change of resistance due to the temperature wave. When these two components are combined there will be a resultant at a frequency 3ω. The potential difference V_3 is observed at two different frequencies, ω_1 and ω_2. The thermal conductivity is given by

$$\lambda = \frac{V^3 \ln(\omega_1/\omega_2)}{4\pi l R^2 (V_{3,1} - V_{3,2})} \frac{dR}{dT},\qquad(3.3)$$

where R is the electrical resistance.

It is noteworthy that, although the 3ω technique involves thermal diffusion, it is essentially the thermal conductivity that is measured.

True thermal diffusivity methods have also been used on thermoelectric materials, particularly at elevated temperatures, since they allow the radiation losses to be eliminated. The thermal diffusivity, κ, is defined as

$$\kappa = \frac{\lambda}{c_V},\qquad(3.4)$$

where c_V is the specific heat per unit volume.

Early thermal diffusivity measurements used the Ångström method [8] in which a sinusoidal variation of temperature is applied at one end of a long bar. The variation of temperature with time is observed at two points separated by a distance l. If there is no heat transfer by radiation it is found that

$$\kappa = \frac{\omega l^2}{2 \ln^2 \alpha} = \frac{\omega l^2}{2\beta^2},$$

(3.5)

where α is the ratio of the amplitudes of the temperature wave at the two points and β is the phase difference. These equations must be modified if there is any radiative transfer but, even when this is a factor, it still remains true that

$$\kappa = \frac{\omega l^2}{2\beta \ln \alpha}.$$

(3.6)

Ångström's method has been used for thermoelectric materials with the thermal wave being generated by the Peltier effect and the specimen acting as its own thermometer [9]. However, it is rare to find a specimen that is long enough for the Ångström conditions to be satisfied. Nevertheless, the measurement of the thermal diffusivity is still favoured as a means of overcoming the problem of radiation losses. It is now more usual to employ thin samples with laser heat sources [10]. Typically, one of the faces is irradiated by a pulsed laser and the temperature at the opposite face is observed as a function of time using, say, an infrared detector. In one technique, the time, $t_{1/2}$, for the back surface to reach half the steady state rise of temperature, is observed. If the thickness is d, the thermal diffusivity is found from the equation

$$\kappa = \frac{1.37d^2}{\pi^2 t_{1/2}}.$$

(3.7)

3.3 The Seebeck coefficient

In some ways the Seebeck coefficient is the easiest of the thermoelectric parameters to measure, since it does not require the use of a sample of any particular shape or size. However, there are dangers to be avoided.

For example, let us suppose that a temperature gradient is established for a particular sample. Then this gradient and the electrical potential gradient can both be determined in principle using fine wire thermocouples. If the thermocouples are made from, say, copper and constantan, the result should be the differential Seebeck coefficient between the test material and copper, assuming that the copper wires are used in the potential measurement. However, there may be a gradient of temperature within the thermocouple junctions and the observed temperature difference may not be equal to that on which the thermoelectric voltage depends.

It is, in fact, not necessary to attach thermocouples or measuring probes to the test specimen. It is sufficient to press copper blocks at different temperatures against the surface of the specimen. Because of the high thermal conductivity of copper, virtually all the temperature gradient will occur in the test material. However, even though this method is so simple it has been found to yield incorrect results. For example, in an attempt to improve the electrical contact, some workers have electroplated the surfaces of their specimens. This has allowed some of the temperature drop to occur in the metallic part of the circuit with a loss of a fraction of the Seebeck voltage.

It is not actually necessary to determine the temperature difference that gives rise to the Seebeck voltage. This is evident from the elegant technique developed by Cowles and Dauncey [11] in which the ratio of the Seebeck coefficient of an unknown material is compared with that of a chromel–alumel thermocouple. Their apparatus is shown in figure 3.7.

The principle of the method is that the ratio of the thermal electromotive force (EMF) from the sample to that from a chromel–alumel couple is equal to the ratio of the calibrated variable resistor, R_2, to the fixed standard resistor, R_1. Balance is first obtained with the switches in the S position by adjustment of the resistor, R_3. Then, balance is again obtained with the switches in the position M by adjustment of R_2. The measurement is completed when balance is reached with the switches in both the S and M positions. In practice, this is a rapid procedure. The Seebeck coefficient is given with respect to chromel. The reversing switch is provided so that materials with both positive and negative Seebeck coefficients can be tested.

Very often it is necessary to determine the Seebeck coefficient as a function of temperature. This can be achieved by maintaining one end of the sample at a fixed temperature while the other end is gradually heated. Continuous measurements are made of the temperature difference and the EMF between the two ends. The Seebeck coefficient is given by the slope of the plot of the EMF against temperature

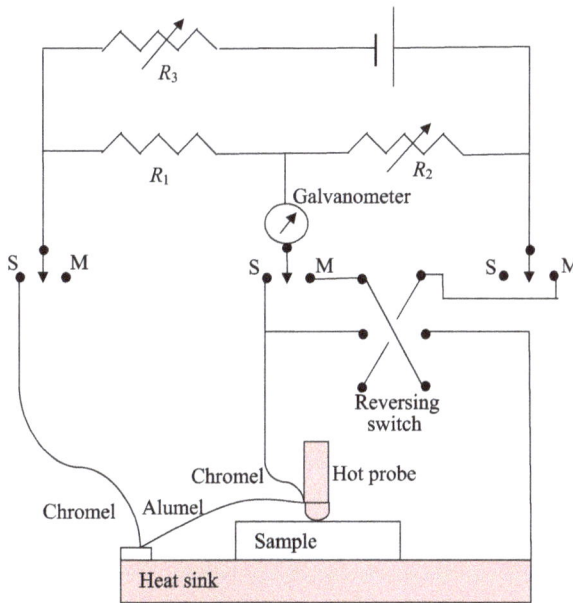

Figure 3.7. Circuit used in the measurement of the Seebeck coefficient [11].

difference. The method is not as accurate as one in which a small temperature difference is applied at each temperature but it is much more rapid.

3.4 Direct determination of the figure of merit

The thermoelectric figure of merit can be calculated from the measured values of the Seebeck coefficient and the electrical and thermal conductivities. However, provided that ZT is not too small, its value can be found directly by comparing the adiabatic and isothermal conductivities, as was first demonstrated by Harman [12].

Suppose that a sample of material is maintained in a vacuum at a uniform temperature and that its electrical resistance is R_I. Then let the electrical resistance be measured again when equilibrium has been reached after the passage of a steady current I. This current produces a temperature difference ΔT through the Peltier effect, where

$$\Delta T = \frac{\alpha T I}{K},\tag{3.8}$$

K being the thermal conductance, it being assumed that all the heat transfer occurs within the sample. There will then be a Seebeck voltage equal to $\alpha \Delta T$ which will be superimposed on the familiar resistive voltage, IR_I. The adiabatic resistance R_A is equal to $R_I + \alpha^2 T/K$ and, thus,

$$\frac{R_A}{R_I} = 1 + \frac{\alpha^2 T}{K} = 1 + ZT.\tag{3.9}$$

We can, therefore, find ZT by measuring the ratio R_A to R_I unless ZT is so small that its determination becomes too inaccurate.

In a typical arrangement, the sample is suspended by its leads in a vacuum, as shown in figure 3.8(a). If, as is usual, all the thermoelectric parameters are sought, the potential probes consist of thermocouples, so that the temperature gradient can be found. Figure 3.9 shows schematically how the potential difference changes with

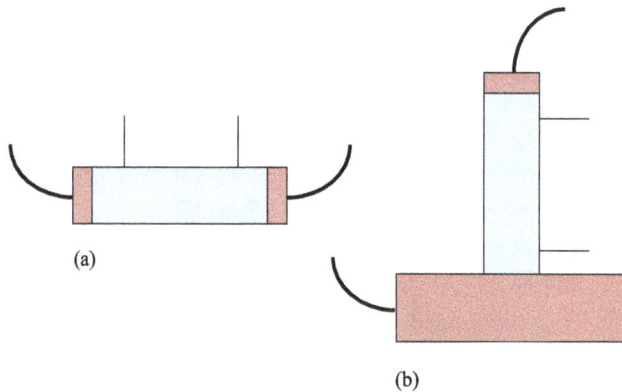

(a)

(b)

Figure 3.8. Configurations for direct measurement of the figure of merit: (a) free sample and (b) sample attached to a heat sink.

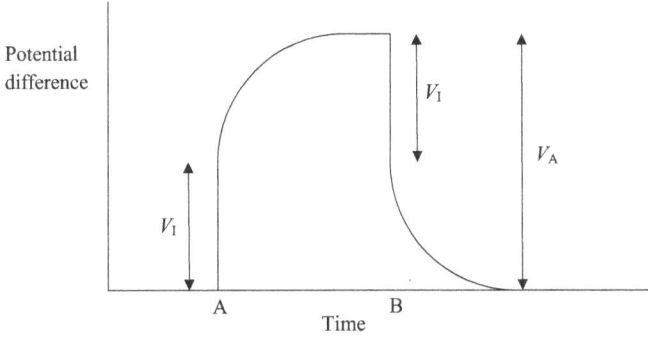

Figure 3.9. Schematic plot of potential difference against time in the measurement of the figure of merit using the Harman method. The current is switched on at A and switched off at B. V_I and V_A are the isothermal and adiabatic voltages, respectively.

time after switching the current on and off. The thermal gradient due to the Peltier effect changes rapidly after switching and some data collection system with a fast response is needed. Alternatively, the EMF can be found using both direct and alternating currents.

It is often convenient to mount the sample on a heat sink, as shown in figure 3.8(b), particularly when the temperature dependence of the figure of merit is being found. However, as we shall see, the losses are greater for this configuration. Often the potential leads are actually attached to the metal blocks at either end of the sample but this is only permissible when it has been established that the electrical contact resistance is negligible.

There are a number of loss terms that have to be accounted for. Thus, there will be conduction of heat through the current leads and the potential probes. There will also be radiation between the surroundings and both the sample and the end caps. However, the Harman technique has the advantage over conventional ways of measuring the thermal conductivity in that there are no heater losses to be considered.

We discuss the losses for the arrangement in figure 3.8(a). The temperature difference ΔT between the ends is taken to be small compared with the absolute temperature. The radiation loss to or from each end cap can be written as $\beta_c A_c \Delta T/2$. Furthermore, the loss by conduction along the lead wires and probes can be expressed as $K_l \Delta T/2$. The radiation loss per unit length from a part of the sample at temperature T will be $\beta P(T - T_0)$, where β is characteristic of the surface of the specimen, P is the perimeter and T_0 is the temperature of the surroundings. On solving the differential equation we find that the dependence of ΔT on the current I can be described by

$$\frac{\alpha I T}{\Delta T} = \frac{\lambda A}{L} + \frac{\beta P L}{12} + \frac{\beta_c A_c}{2} + \frac{K_l}{2}. \tag{3.10}$$

The first term on the right-hand side of this equation corresponds to the heat conducted through the sample while the remaining terms are related to the losses.

The loss terms are somewhat larger if the experimental arrangement in figure 3.8(b) is used. In this case equation (3.10) becomes

$$\frac{\alpha IT}{\Delta T} = \frac{\lambda A}{L} + \frac{\beta PL}{3} + \beta_c A_c + K_l.$$ (3.11)

In either version of the method the losses can be determined by using samples of different length and shape.

It is, perhaps, surprising that no mention has been made of the influence of Joule heating. In fact, it can be shown [13] that the Joule effect certainly raises the average temperature of the measurement but it does not otherwise affect the value of ZT.

There is, however, another unexpected source of error. This can arise when the sample is inhomogeneous [14]. An extreme case of this source of error is found for a sample that is made up of equal parts of materials with positive and negative Seebeck coefficients. In this situation the average Seebeck coefficient may be close to zero but there will still be localised Peltier heating and cooling. Any attempt to determine, say, the thermal conductivity by the Harman method will then lead to a substantial error. Indeed there will be an error for any sample in which the Seebeck coefficient is not uniform. Nevertheless, the Harman technique remains one of the most useful methods for testing thermoelectric materials.

References

[1] Putley E H 1955 *Proc. Phys. Soc.* B **68** 35
[2] Goldsmid H J 1958 *Proc. Phys. Soc.* **71** 633
[3] Dauphinee T M and Woods S B 1955 *Rev. Sci. Instrum.* **26** 693
[4] Ioffe A V and Ioffe A F 1958 *Sov. Phys. Tech. Phys.* **3** 2163
[5] Kaganov M A 1958 *Sov. Phys. Tech. Phys.* **3** 2169
[6] Swann W F G 1959 *J. Franklin Inst.* **267** 363
[7] Cahill D G 1990 *Rev. Sci. Instrum.* **61** 802
[8] Ångström A J 1861 *Ann. Phys.* **114** 513
[9] Green A and Cowles L E J 1960 *J. Sci. Instrum.* **37** 349
[10] Tritt T M and Weston D 2004 *Thermal Conductivity: Theory, Properties and Applications* ed T M Tritt (New York: Kluwer/Plenum) p 197
[11] Cowles L E J and Dauncey L A 1962 *J. Sci. Instrum.* **39** 16
[12] Harman T C 1958 *J. Appl. Phys.* **29** 1373
[13] Nolas G S, Sharp J and Goldsmid H J 2001 *Thermoelectrics: Basic Principles and New Materials Developments* (Berlin: Springer) p 102
[14] Goldsmid H J 2006 *J. Thermoelectricity*, No. 1 5

The Physics of Thermoelectric Energy Conversion

H Julian Goldsmid

Chapter 4

Electronic transport in semiconductors

4.1 Energy band theory

The treatment of charge transport in solids is derived from the free electron theory of metals. Because of the interaction between the electrons and the crystal lattice, the electrons are confined to states in specific bands of energy that are separated by energy gaps. Moreover, the probability that an energy state contains an electron is governed by quantum mechanics rather than classical statistics. It is found, at least near the edges of the energy bands, that the carriers behave like free electrons except that they must be assigned an effective mass m^* that is different from the free electron mass, m.

For our purposes, we may confine our attention to the two bands of highest energy, the conduction and valence bands. Electrons can take part in the conduction process only if they reside in energy states that are close to vacant states. This means that conduction in a metal is due to electrons having a narrow range of energies. There is a particular energy, the Fermi energy, at which there is a 50% probability of a state being filled. It is the states within a few kT of the Fermi level, then, that are responsible for the transport phenomena.

We are most interested in materials in which the Fermi level lies close to the edge of a band. If this is the conduction band, the carriers may be regarded as quasi-free electrons. On the other hand, if the Fermi level lies close to the upper edge of the next-lower band, the valence band, the electrons behave as if their effective mass is negative. It is convenient to regard these carriers in the valence band as if they have a positive mass and a positive charge, and they are commonly known as positive holes (or just as holes).

The density of electron states is much smaller near the band edge than it is deep within a band. Thus, when conduction is due to carriers near the band edges, the conductivity is much less than it is for a metal and the substance is known as a semiconductor. Semiconductors are called n-type or p-type according to whether the conduction is primarily due to electrons or holes. It is noted that n-type and p-type

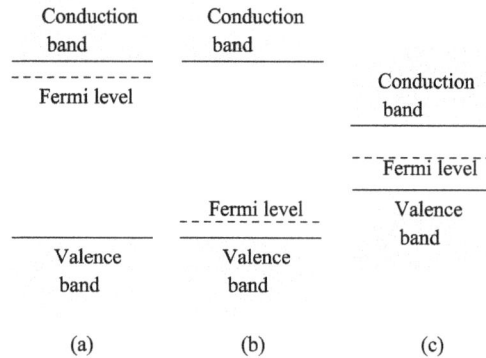

Figure 4.1. Energy band diagrams for semiconductors: (a) n-type, (b) p-type and (c) intrinsic.

semiconductors have negative and positive Seebeck coefficients respectively. Typical energy diagrams for n- and p-type semiconductors are shown in figure 4.1(a) and (b), respectively. Figure 4.1(c) shows the case of an intrinsic semiconductor with the Fermi level close to the middle of the energy gap.

For the time being we consider semiconductors in which the energy difference between the valence and conduction bands is large enough for only one type of carrier to be significant. The electrical conductivity and the Seebeck coefficient will then depend on the Fermi energy, as measured from the band edge, the effective mass, in so far as it determines the density of states in the band, and a quantity known as the carrier mobility, μ, which is defined as the drift speed of the carriers in unit electric field.

4.2 Mobility and effective mass

The energy diagrams in figure 4.1 do not show the whole picture. An electronic state is characterised not only by its energy but also by its wave vector. In the simplest case, the energy minimum occurs at zero wave vector. However, a minimum may also be found at a non-zero value for the wave vector, crystal symmetry requiring that similar minima are located at corresponding points in wave vector space. The material is then called a multi-valley conductor.

In the quasi-free electron theory, the density of electron states at an energy E is given by

$$g(E)\,\mathrm{d}E = \frac{4\pi(2m^*)^{3/2}E^{1/2}\,\mathrm{d}E}{h^3}, \tag{4.1}$$

where h is Planck's constant. This equation must be modified in an N_V-valley conductor by setting m^* equal to $N_V^{2/3}$ times its value for a single valley. The single-valley effective mass may exhibit directional dependence and is termed the inertial mass m_I to distinguish it from the density-of-states mass, m^*.

The mobility depends on the inertial effective mass and on the relaxation time associated with the scattering processes. It is assumed that any disturbance in the carrier distribution will relax towards its equilibrium value with a characteristic time, τ, which it will be supposed is dependent on the energy, E, and may be written as $\tau_0 E^r$, where r depends on the scattering process. It seems that for many of the materials in which we are interested acoustic-mode lattice scattering is predominant, in which case r is equal to $-1/2$. This is in spite of the fact that the bonding may have an ionic as well as a covalent component.

4.3 Dependence of the transport properties on the Fermi energy

The density of charge carriers in any particular metal is more-or-less a fixed quantity but this is not so for a semiconductor. The carrier density and, indeed, the sign of the majority carriers can be controlled by the addition of impurities. Impurities that increase the electron concentration are called donors and those that increase the hole concentration are known as acceptors. An intrinsic semiconductor contains either no impurities or an equal number of donors and acceptors. In some materials, vacancies on lattice sites can act in the same way as foreign atoms. One of the effects of these impurities is to change the position of the Fermi level. We shall find it convenient to regard the Fermi energy, E_F, as the independent variable.

We make use of the Boltzmann equation that relates the disturbance in the electron distribution to the applied electric field and temperature gradient. It is supposed that this disturbance is small and that it relaxes towards the equilibrium distribution according to the equation

$$\frac{f(E) - f_0(E)}{\tau} = u\frac{df_0(E)}{dE}\left[\frac{dE_F}{dx} + \frac{(E - E_F)}{T}\frac{dT}{dx}\right], \tag{4.2}$$

where u is the velocity of the carriers in the x direction and $f(E)$ is the Fermi distribution function that has the equilibrium value $f_0(E)$ given by

$$f_0(E) = \frac{1}{\exp\left(\frac{E - E_F}{kT}\right) + 1}. \tag{4.3}$$

The transport properties can be found from the relations between the gradients of the electric potential and temperature, the electric current density, i, and the heat flux density, j. The expressions for the electric current and heat flux densities are

$$i = \mp\int_0^\infty euf(E)g(E)dE, \tag{4.4}$$

and

$$j = \int_0^\infty u(E - E_F)f(E)g(E)dE. \tag{4.5}$$

In equation (4.4) the upper sign is applicable when the carriers are electrons and the lower sign applies for hole conduction. In the latter case the energy is measured downwards from the band edge.

The electrical conductivity is found by setting the temperature gradient equal to zero, while the Seebeck coefficient and electronic thermal conductivity require the electric current to be zero. In solving these equations, it may be assumed that the disturbance in the distribution of the carriers is small enough that any alteration in u is much less than the equilibrium velocity. We also replace the disturbed distribution function $f(E)$ by $f(E) - f_0(E)$ since the electrical and thermal flows are zero in equilibrium.

It is convenient to express the transport coefficients in terms of integrals defined as

$$K_s = -\frac{2T}{3m^*} \int_0^\infty g(E)\tau_e E^{s+1}\frac{\mathrm{d}f_0(E)}{\mathrm{d}E}\,\mathrm{d}E. \tag{4.6}$$

This expression, in turn, may be written as

$$K_s = \frac{8\pi}{3}\left(\frac{2}{h^2}\right)^{3/2}(m^*)^{1/2}T\tau_0\left(s + r + \frac{3}{2}\right)(kT)^{s+r+3/2}F_{s+r+1/2}, \tag{4.7}$$

where

$$F_n(\xi) = \int_0^\infty \xi^n f_0(\xi)\,\mathrm{d}\xi. \tag{4.8}$$

Here ξ is the reduced energy, E/kT. We shall also use the symbol η to represent the reduced Fermi energy, E_F/kT. The values of F are known as the Fermi–Dirac integrals.

It is found that the electronic parameters that appear in the figure of merit are

$$\sigma = \frac{e^2}{T}K_1, \tag{4.9}$$

$$\alpha = \pm\frac{1}{eT}\left(E_F - \frac{K_1}{K_0}\right), \tag{4.10}$$

and

$$\lambda_e = \frac{1}{T^2}\left(K_2 - \frac{K_1^2}{K_0}\right). \tag{4.11}$$

It must be remembered that the total thermal conductivity, λ, is the sum of the electronic component given by equation (4.11) and a lattice component, λ_L.

4.4 Degenerate and non-degenerate conductors

There are good approximations for the Fermi–Dirac integrals when the Fermi energy is either very much less than or very much greater than zero. If $E_F > 4kT$ the material is said to be degenerate and the metallic approximation is used. In this case

$$F_n(\eta) = \frac{\eta^{n+1}}{n + 1} + n\eta^{n-1}\frac{\pi^2}{6} + n(n - 1)(n - 2)\eta^{n-3}\frac{7\pi^4}{360} + \cdots. \tag{4.12}$$

One must include as many terms as are needed for the parameter in question to have a non-zero value. Thus, the electrical conductivity requires only the first term so that

$$\sigma = \frac{8\pi}{3}\left(\frac{2}{h^2}\right)^{3/2} e^2(m^*)^{1/2}\tau_0 E_F^{r+3/2}. \tag{4.13}$$

The electronic thermal conductivity needs the first two terms on the right-hand side of equation (4.12) to be included. Then it is found that

$$\frac{\lambda_e}{\sigma T} = \frac{\pi^2}{3}\left(\frac{k}{e}\right)^2. \tag{4.14}$$

For most metals the electronic thermal conductivity is much larger than the lattice contribution. Thus, equation (4.14) embodies the Wiedemann–Franz law which states that the ratio of the thermal conductivity to the electrical conductivity is the same for all metals, at any given temperature. The ratio $\lambda_e/\sigma T$ is known as the Lorenz number, L.

The same two terms in equation (4.12) are also needed for the Seebeck coefficient. It is found that

$$\alpha = \mp\frac{\pi^2}{3}\frac{k}{e}\frac{\left(r + \frac{3}{2}\right)}{\eta}. \tag{4.15}$$

It is clear that, as η becomes large, the Seebeck coefficient has a magnitude that is much less than k/e, which is consistent with the fact that most metals have values of α of the order of only a few $\mu V\ K^{-1}$.

We are actually much more interested in materials for which η is close to zero or negative. When η is less than -2 we may use the classical approximation in which the Femi–Dirac integrals become

$$F_n(\eta) = \exp(\eta)\Gamma(n + 1). \tag{4.16}$$

The gamma function Γ is such that $\Gamma(n + 1)$ is equal to $n\Gamma(n)$. When n is an integer, $\Gamma(n + 1)$ is equal to $n!$ and $\Gamma(1/2)$ is equal to $\pi^{1/2}$. Thus, we can easily calculate the gamma function for both integral and half-integral values of n.

If we use the classical approximation the integrals K_s become

$$K_s = \frac{8\pi}{3}\left(\frac{2}{h^2}\right)^{3/2} (m^*)^{1/2}T\tau_0(kT)^{s+r+3/2}\Gamma\left(s + r + \frac{5}{2}\right)\exp(\eta). \tag{4.17}$$

Then the Seebeck coefficient is

$$\alpha = \mp\frac{k}{e}\left[\eta - \left(r + \frac{5}{2}\right)\right]. \tag{4.18}$$

It will be seen that the Seebeck and Peltier coefficients are a measure of the total energy transported by the charge carriers weighted according to the scattering parameter, r. It is noted that the range for which this equation is valid covers Seebeck coefficients of greater magnitude than $4k/e$, if we suppose that r is $-1/2$.

Since k/e is 86.4 μV K^{-1}, this means that the magnitude of the Seebeck coefficient should exceed about 350 μV K^{-1} if classical statistics are to apply. For most of the thermoelectric materials that are used today, the Seebeck coefficient has a smaller value than this, so the classical condition cannot often be used, except as a gross approximation.

In the classical range the electrical conductivity is given by

$$\sigma = \frac{8\pi}{3}\left(\frac{2}{h^2}\right)^{3/2} e^2(m^*)^{1/2}\tau_0(kT)^{r+3/2}\Gamma\left(r + \frac{5}{2}\right)\exp(\eta). \tag{4.19}$$

It is common practice to express the electrical conductivity as

$$\sigma = ne\mu, \tag{4.20}$$

where n is the carrier concentration and μ is the mobility. The expression for the carrier concentration is

$$n = 2\left(\frac{2\pi m^* kT}{h^2}\right)^{3/2} \exp(\eta), \tag{4.21}$$

where the quantity $2(2\pi m^* kT/h^2)^{3/2}$ is known as the effective density of states. If we substitute the carrier concentration in equation (4.19) we find that the mobility is given by

$$\mu = \frac{4}{3\pi^{1/2}}\Gamma\left(r + \frac{5}{2}\right)\frac{e\tau_0(kT)^r}{m^*}. \tag{4.22}$$

It is noteworthy that the mobility does not depend directly on the Fermi energy in the classical region.

The expression for the Lorenz number in a non-degenerate conductor is

$$L = \left(\frac{k}{e}\right)^2\left(r + \frac{5}{2}\right), \tag{4.23}$$

which is of the same order as the value given by equation (4.14) for a metal, though somewhat smaller.

Although it is better to use the classical rather than the degenerate approximation for most thermoelectric materials, neither is really applicable. Thus, one must generally use the full expressions for the Fermi–Dirac integrals, F_n. Tables of these integrals for integral and half-integral values of n may be found elsewhere [1–3].

4.5 Optimising the Seebeck coefficient

If we were restricted to metallic conductors the figure of merit would rise continuously with the Seebeck coefficient. This is because the ratio of electrical to thermal conductivity would always have the same value. However, in reality, as the carrier concentration falls, the thermal conductivity becomes greater than the value expected from the Wiedemann–Franz law. This is because of the influence of heat conduction by the lattice. We shall discuss the lattice conductivity in the next chapter but here we take note of its existence since it affects the preferred value for the Seebeck coefficient.

If the lattice conductivity were very large compared with the electronic thermal conductivity, as it is for many semiconductors, the figure of merit would be proportional to a quantity known as the power factor, which is defined as $\alpha^2 \sigma$. As shown in figure 4.2, the power factor falls off slowly as the Fermi level moves into the band and the Seebeck coefficient decreases. It also falls rapidly as the Fermi level moves into the band gap due to decrease in the carrier concentration. The maximum power factor occurs when the Fermi level is very close to the band edge.

When we take account of the lattice conductivity in calculating the figure of merit it is clear that the optimum Fermi energy will become more negative than that for the maximum power factor. This is apparent from the curves shown in figure 4.3. Here the dimensionless figure of merit is plotted against the reduced Fermi energy for different values of $(zT)_{max}$. As $(zT)_{max}$ becomes larger, so also does the optimum Fermi level move further into the energy gap. This means that the optimum Seebeck coefficient becomes of greater magnitude, as shown in figure 4.4.

The value of zT for any particular Fermi energy depends on the carrier mobility, the density-of-states effective mass and the lattice thermal conductivity. These three parameters can be embodied in a single quantity β which is given by [4]

$$\beta = \left(\frac{k}{e}\right)^2 \frac{\sigma_0 T}{\lambda_L}, \tag{4.24}$$

where

$$\sigma_0 = 2e\mu \left(\frac{2\pi m^* k T}{h^2}\right)^{3/2}. \tag{4.25}$$

It may be noted that $(zT)_{max}$ reaches a value of about 1 when β is equal to 0.4.

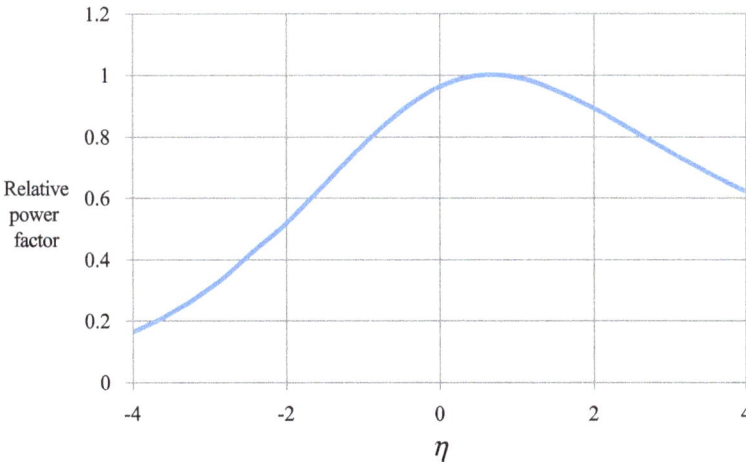

Figure 4.2. Plot of power factor against reduced Fermi energy for $r = -1/2$. The power factor is expressed as a fraction of its maximum value.

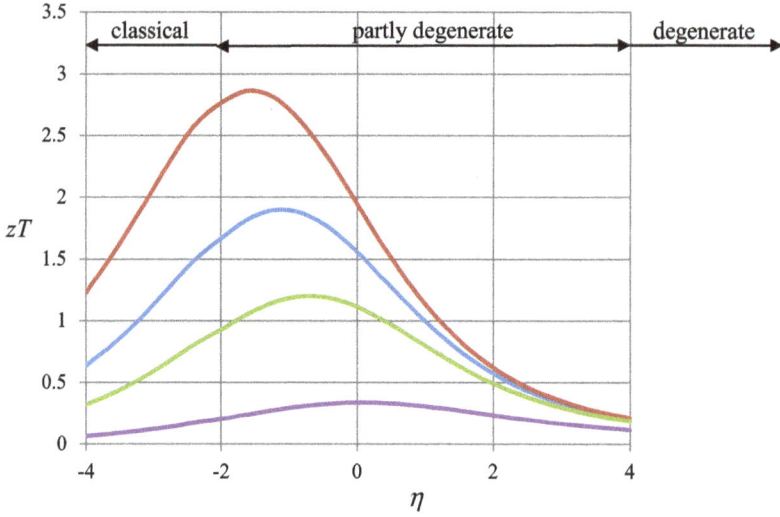

Figure 4.3. Plots of zT against reduced Fermi energy for various values of $(zT)_{max}$.

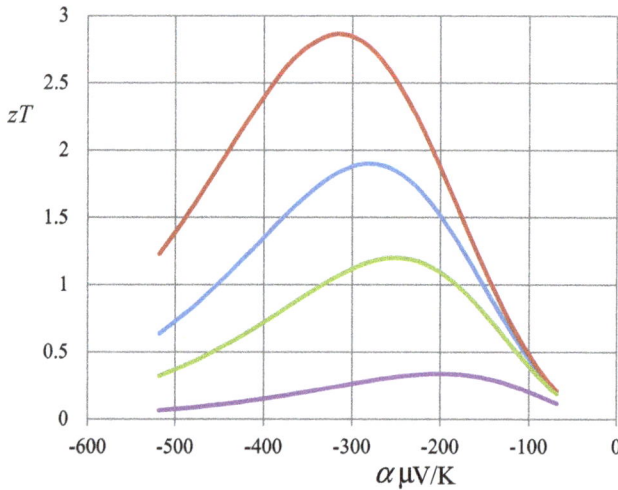

Figure 4.4. Plots of zT against Seebeck coefficient for various values of $(zT)_{max}$. The graphs are for n-type material but would be identical for p-type material apart from the sign of the Seebeck coefficient.

4.6 Bipolar conduction

As the temperature is raised it becomes possible for electrons to be thermally activated from the valence band to the conduction band. The concentration of electron–hole pairs depends on the size of the energy gap, E_g. Provided that E_g is small enough, these carriers may become more numerous than those due to donor or acceptor impurities. With increase of temperature the conductor passes from the extrinsic region into the region of mixed conduction and finally becomes intrinsic. The presence of both electrons and holes in the same conductor can have a profound effect on the thermoelectric properties.

There will be contributions i_n and i_p to the electric current density from the electrons and holes respectively. These contributions satisfy the equation

$$i_{n,p} = \sigma_{n,p}\left(\frac{dV}{dx} - \alpha_{n,p}\frac{dT}{dx}\right),\tag{4.26}$$

where $\sigma_{n,p}$ and $\alpha_{n,p}$ are the partial conductivities and Seebeck coefficients. The electrical conductivity is found by setting dT/dx equal to zero and, not surprisingly, it has the value

$$\sigma = \sigma_n + \sigma_p.\tag{4.27}$$

The Seebeck coefficient is obtained when we set $i_n + i_p$ equal to zero, whence

$$\alpha = \frac{\alpha_n\sigma_n + \alpha_p\sigma_p}{\sigma_n + \sigma_p}.\tag{4.28}$$

This equation tells us that the overall Seebeck coefficient is a weighted average of the partial Seebeck coefficients, which will be of opposite sign. This means that the Seebeck coefficient of a mixed or intrinsic semiconductor is likely to be very small.

There is a remarkable result if we determine the electronic thermal conductivity when both types of carrier are present. Then the heat flux densities for the two carriers are given by

$$j_{n,p} = \alpha_{n,p}Ti_{n,p} - \lambda_{n,p}\frac{dT}{dx}.\tag{4.29}$$

The thermal conductivity is defined for the condition of zero total electric current. Thus,

$$\lambda_e = \lambda_n + \lambda_p + \frac{\sigma_n\sigma_p}{\sigma_n + \sigma_p}(\alpha_n - \alpha_p)^2 T.\tag{4.30}$$

The third term on the right-hand side of equation (4.30) is the contribution to the thermal conductivity from the bipolar effect and may be an order of magnitude greater than the partial conductivities of the single carriers [5].

It is concluded that mixed conduction should be avoided in thermoelectric materials since it not only reduces the Seebeck coefficient but also increases the thermal conductivity.

4.7 Band engineering and nanostructure effects

Here we discuss some of the ways in which the power factor for a given Fermi energy might be improved.

It is evident from equation (4.10) and, particularly, its classical form, equation (4.18), that the Seebeck coefficient has a potential energy component and a contribution from the kinetic energy. The kinetic energy component is weighted according to the form of scattering for the charge carriers. In most high mobility semiconductors the scattering parameter, r, is equal to $-1/2$ and the relaxation time is greatest for the carriers of the lowest energy. On the other hand, if ionized-impurity scattering becomes dominant, r rises to $+3/2$, and the high-energy carriers are the least strongly scattered. There is then a substantial rise in the kinetic energy that is transported by the charge carriers. Of course, this is accompanied by a decrease in the mobility but, as shown by Ioffe [6], the overall effect could be advantageous. In practice it appears that this effect has never been used to advantage. It would seem to be most beneficial in semiconductors with narrow energy gaps since then the potential energy of either type of carrier cannot be equal to more than about half the gap.

Another way of improving the power factor involves the introduction of additional energy states. This may come about through the addition of specific impurities that give rise to states above the edge of the main band. There is also the possibility of selecting materials in which there are additional bands with edges not too far removed from the edge of the original band.

It was proposed by Hicks and Dresselhaus [7] that it might be advantageous to make use of nanostructured semiconductors. Nanostructures can be two-dimensional in the form of thin sheets, one-dimensional as nanowires or nanotubes, or even zero-dimensional as nanodots. In all cases, the band structure becomes modified.

Following the theory of Hicks and Dresselhaus we consider the case of a conduction band with a parabolic density of states. The simplest situation is that of a two-dimensional sheet of thickness d that is of the order of a few interatomic spacings. This means that the dispersion relation is changed from

$$E = \frac{\hbar^2 k_x^2}{2m_x} + \frac{\hbar^2 k_y^2}{2m_y} + \frac{\hbar^2 k_z^2}{2m_z}, \tag{4.31}$$

to

$$E = \frac{\hbar^2 k_x^2}{2m_x} + \frac{\hbar^2 k_y^2}{2m_y} + \frac{\hbar^2 \pi^2}{2m_z d^2}. \tag{4.32}$$

We introduce a quantity η^* which is related to the reduced Fermi energy η by the relation

$$\eta^* = \eta - \frac{\hbar^2 \pi^2}{2m_x d^2 kT}. \tag{4.33}$$

In terms of this quantity the Seebeck coefficient is given by

$$\alpha = -\frac{k}{e}\left(\frac{2F_1}{F_0} - \eta^*\right), \tag{4.34}$$

where the relaxation time has been supposed to be constant. The electrical conductivity is

$$\sigma = \frac{1}{2\pi d}\left(\frac{2kT}{\hbar^2}\right)(m_x m_y)^{1/2} F_0 e \mu_x. \tag{4.35}$$

When the expression for the electronic thermal conductivity is also included, the dimensionless figure of merit becomes

$$zT = \frac{\left(2\frac{F_1}{F_0} - \eta^*\right)^2 F_0}{\frac{1}{\beta^*} + 3F_2 - 4\frac{F_1^2}{F_0}}. \tag{4.36}$$

where β^* is given by

$$\beta^* = \frac{1}{2\pi d}\left(\frac{2kT}{\hbar^2}\right)(m_x m_y)^{1/2}\frac{k^2 T \mu_x}{e\lambda_L}. \tag{4.37}$$

Equation (4.36) does not become significantly different from its three-dimensional equivalent until d is very small. Eventually, when d is small enough, the effective density of states rises and this allows zT to become greater.

Hicks and Dresselhaus applied their ideas to the most widely used thermoelectric material, bismuth telluride. They assumed a rather moderate maximum value of zT equal to 0.52 for the bulk compound and ignored any possible change in the lattice conductivity. They predicted that zT would become substantially greater than unity for specimens of less than 5 nm thickness. An even greater improvement would be expected for one-dimensional materials and quantum dots and similar behaviour has been predicted for other substances.

In the event, it seems that there are very few instances where it can be claimed that the electronic properties have been enhanced by adopting nanostructures. This is possibly due to the difficulty in dealing with materials that have a sufficiently small thickness. Nevertheless, there are numerous examples of materials that have been improved through incorporating nanostructures but this has usually been attributed to a reduction in the lattice conductivity.

References

[1] McDougall J and Stoner E C 1938 *Phil. Trans. Roy. Soc. Lond.* A **237** 67
[2] Rhodes P 1950 *Proc. Roy. Soc. Lond.* A **204** 396
[3] Goldsmid H J 2016 *Introduction to Thermoelectricity, Second Edition* (Berlin, Heidelberg: Springer) pp 31–2
[4] Chasmar R P and Stratton R 1959 *J. Electron. Control* **7** 52
[5] Price P J 1955 *Phil. Mag.* **46** 1252
[6] Ioffe A F 1957 *Semiconductor Thermoelements and Thermoelectric Cooling* (London: Infosearch) p 146
[7] Hicks L D and Dresselhaus M S 1993 *Phys. Rev.* B **47** 12727

Chapter 5

Heat conduction by the crystal lattice

5.1 Phonon conduction in pure crystals

The conduction of heat occurs in all solids whether or not they are electrical conductors. Thermal energy is contained within the lattice vibrations and these are responsible for the transport of heat. It is interesting to note that, although metals in general are good heat conductors, the material with the highest thermal conductivity is an electrical insulator, pure diamond.

When heat is supplied to a solid, the energy within the vibrational modes rises. The heat that is needed to raise the temperature by one degree is the specific heat, which, for our purposes may be expressed per unit volume, with the value c_V. A reasonably accurate, if over simplified, theory of the specific heat was provided by Debye.

In the Debye theory, the material is taken to be in the form of an elastic continuum. The dispersion curve, or the plot of frequency, ν, against wave number, q, is linear as shown in figure 5.1(a). The velocity of the vibrational waves is given by the slope of the plot and is constant. The total number of modes is equal to the number of atoms, this setting a higher limit on the frequency and wave number.

The atomistic nature of a crystal has an effect on the dispersion curve, particularly at the highest frequencies. This effect is shown in figure 5.1(b), which shows schematically the situation for a linear chain. It has been assumed that the chain consists alternately of atoms of different mass, in which case there are both acoustic vibrations, where adjacent atoms are moving in the same direction, and optical vibrations, where they can move apart. The diagram shows the acoustic branch and an optical branch. If there are n atoms in the unit cell of a three-dimensional crystal, then there will be three acoustic branches and $3(n - 1)$ optical branches, taking account of both longitudinal and transverse waves.

When we are discussing the conduction of heat it is important to note that the group velocity is given by the slope of the dispersion curve and can be very small for the optical modes.

doi:10.1088/978-1-6817-4641-8ch5

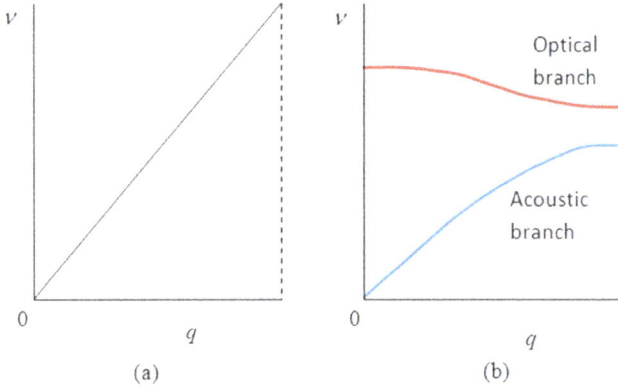

Figure 5.1. Schematic dispersion curves for (a) the Debye continuum and (b) a diatomic linear chain.

The vibrational modes are distributed in wave vector space up to a total determined by the number of atoms. In Debye's theory, the number of modes varies as the square of the frequency but, in reality, the vibrational spectrum may be much more complex. Nevertheless, the Debye model is satisfactory for low frequencies and, as it happens, the specific heat is not very sensitive to the details of the spectrum.

The specific heat is determined from the energy contained within the vibrational modes and the rate of change of this energy with temperature. The occupation of these modes is determined by Bose–Einstein statistics for which the energy in a mode of frequency ν is $h\nu\{\exp(h\nu/kT) - 1\}^{-1}$. Using the Debye model it is found that the specific heat is proportional to $(T/\theta_D)^3 F_D(\theta_D/T)$, where θ_D is known as the Debye temperature and F_D is given by

$$F_D\left(\frac{\theta_D}{T}\right) = \int_0^{\theta_D/T} \frac{\left(\frac{h\nu}{kT}\right)^4 \exp\left(\frac{h\nu}{kT}\right)}{\left\{\exp\left(\frac{h\nu}{kT}\right) - 1\right\}^2} \mathrm{d}\left(\frac{h\nu}{kT}\right). \tag{5.1}$$

At very low temperatures the Debye theory predicts that the specific heat should vary as T^3 and indeed this is observed in practice, though at rather lower temperatures than expected. When T becomes greater than θ_D the specific heat should become independent of temperature as also agrees with experiment. However, observations show that somewhat different values of θ_D are needed to fit the results at different temperatures so this quantity cannot really be regarded as a constant, though it is still a useful parameter.

Debye [1] attempted to use his theory to explain Eucken's law [2], which states that the thermal conductivity of a pure non-metallic crystal varies inversely with the absolute temperature. He showed that this law would follow if the lattice vibrational waves scattered one another. His difficulty, though, was that there should be no scattering between perfectly harmonic waves. This difficulty was solved by Peierls [3], who introduced the idea of phonons, or quantised lattice vibrations.

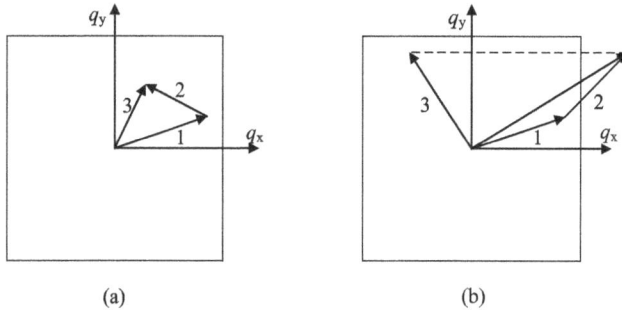

Figure 5.2. The difference between a normal process (a) and an umklapp process (b). The diagrams show a unit cell in wave vector space for a square lattice.

Peierls showed that phonons can scatter one another in two ways. In what is called a normal process, two phonons, q_1 and q_2 interact to produce a third phonon q_3 which lies within the same unit cell in wave vector space. However, in an umklapp process, the resultant of the addition of q_1 and q_2 is a phonon with a wave vector outside this unit cell, which can be brought back into the cell by the addition of a reciprocal lattice vector, that is a vector with the width of the unit cell. The processes are illustrated in figure 5.2 for a simple two-dimensional crystal with a square lattice.

It is the umklapp processes that are responsible for the lattice conductivity being finite. They are a consequence of an anharmonic component in the lattice vibrations. Peierls showed that their probability increases with rise of temperature and their effect is consistent with Eucken's law.

At very low temperatures nearly all the phonons have very small wave vectors. Also, it is apparent that umklapp processes cannot occur unless at least one of the interacting phonons has a wave vector equal to at least half the width of the reciprocal lattice cell, $q_{max}/2$. In fact, since the probability of any phonon having a wave vector much greater than $q_{max}/2$ is exceedingly low, it is likely that both the phonons involved in an umklapp process will have wave vectors close to $q_{max}/2$. On this basis, Peierls was able to show that the mean free path of phonons at low temperatures should be proportional to $\{\exp(-\theta_D/aT)\}^{-1}$, where a lies close to 2. In fact, it was found that the Eucken law persisted to lower than expected temperatures and the exponential behaviour that resulted from the calculations of Peierls was not immediately confirmed. Later it was realised that scattering of phonons on various defects was masking the exponential effect.

Although the normal processes are momentum-conserving, it must not be thought that they have no influence on the thermal resistance. They do act so as to redistribute the energy within the phonon system. Thus, they may sometimes lead to an increase in the number of phonons that can be scattered by other processes.

5.2 Prediction of the lattice conductivity

It is not easy to make an accurate prediction of the lattice conductivity of any pure crystal. However, a useful approximate expression was derived by Keyes [4] and we shall discuss his theory here.

We start with the well-established equation

$$\lambda_L = \frac{1}{3} c_V v l_t, \tag{5.2}$$

where the specific heat c_V is defined for unit volume, v is the speed of sound and l_t is the phonon mean free path. Dugdale and MacDonald [5] suggested that the lattice conductivity should bear some relationship to the thermal expansion coefficient, α_T, since both quantities are dependent on the anharmonicity. They proposed that the anharmonicity should be represented by the dimensionless quantity, $\alpha_T \gamma T$, where γ is the Grüneisen parameter. They then suggested that the mean free path of the phonons should be close to $a/\alpha_T \gamma T$ where a is the lattice constant. Thus, from equation (5.2) we find that

$$\lambda_L = \frac{c_V a v}{3 \alpha_T \gamma T}. \tag{5.3}$$

At this stage we introduce the Debye equation of state, which relates the expansion coefficient to the compressibility, χ. Then,

$$\alpha_T = \frac{\chi \gamma c_V}{3}. \tag{5.4}$$

Also, the speed of sound can be expressed in terms of the Debye temperature through the relation

$$v = (\rho \chi)^{-1/2} = \frac{2 k a \theta_D}{h}, \tag{5.5}$$

where ρ is the density. Then, putting the atomic volume, V, equal to the cube of the lattice constant, we obtain the equation

$$\lambda_L = 8 \left(\frac{k}{h} \right)^3 \frac{M V^{1/3} \theta_D^3}{\gamma^2 T}, \tag{5.6}$$

where M is the atomic mass. It must be stressed that several approximations have been used in the derivation of equation (5.6) but it is nevertheless a useful tool.

By making use of further approximate relationships we may obtain an equation for λ_L that involves three easily determined properties, namely the melting temperature, T_m, the density and the mean atomic weight, A. To this end we use equations (5.5) and (5.6) to obtain Lawson's equation [6]

$$\lambda_L = \frac{a}{3 \gamma^2 T \chi^{3/2} \rho^{1/2}}. \tag{5.7}$$

We also use the Lindemann melting rule [7]

$$T_m = \frac{\varepsilon_m V}{R \chi}, \tag{5.8}$$

where T_m is the melting temperature and R is the gas constant. This rule is based on the approximation that all solids melt when the atomic vibrations are large enough to reach a certain fraction, ε_m, of the lattice constant, ε_m having a universal value. Thence we finally obtain the Keyes relation

$$\lambda_L T = B_K \frac{T_m^{3/2} \rho^{2/3}}{A^{7/6}}, \tag{5.9}$$

where

$$B_K = \frac{R^{3/2}}{3\gamma^2 \varepsilon_m^3 N_A^{1/3}}, \tag{5.10}$$

and N_A is Avogadro's number. It is noted that B_K contains the quantities ε_m and γ that do not change much from one material to another. The Keyes relation is consistent with the observation that the lattice conductivity is inversely proportional to the temperature for pure crystals.

The Keyes relation suggests that the lattice conductivity should fall as the mean atomic weight rises. This decrease may be offset to some extent by a rise in the density, but the melting temperature is likely to become smaller as the atomic weight becomes greater. It was observed by Ioffe and Ioffe [8] that the lattice conductivity falls with increasing mean atomic weight for a number of systems, as shown in figure 5.3. The elementary group IV semiconductors generally have high lattice conductivities with somewhat lower values for the III–V compounds. The alkali halides have a much lower range of lattice conductivities, particularly when the atomic weight ratio is large. A small value for λ_L is, of course, most desirable for a thermoelectric material

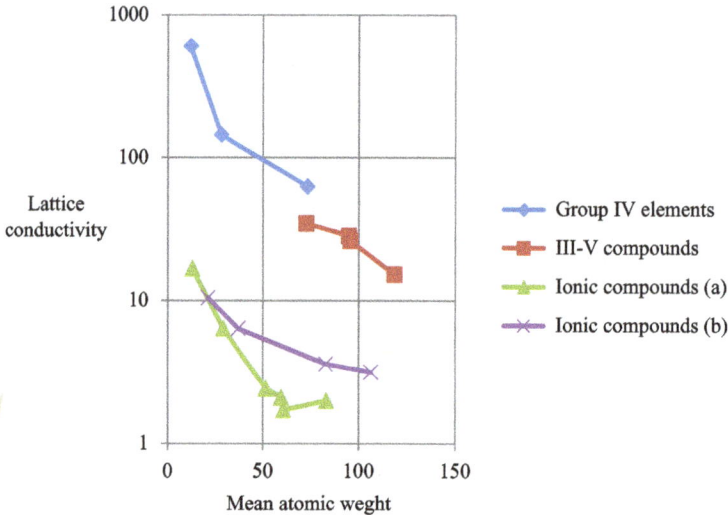

Figure 5.3. Variation of lattice conductivity with mean atomic weight for different systems as observed by Ioffe and Ioffe [8]. The ionic compounds are alkali halides with (a) atomic weight ratios greater than 1.5 and (b) less than 1.5.

but a high mobility for the charge carriers is equally desirable and this rules out some of the substances that have small lattice conductivities.

5.3 Solid solutions

There are a number of reasons why one might wish to use a solid solution between two semiconductors to improve the thermoelectric properties. For example, it might be necessary to increase the energy gap of a given material so as to raise the operating temperature without the onset of bipolar conduction. However, it was Ioffe and his colleagues [9] who first suggested the use of solid solutions to reduce the lattice conductivity. By solid solutions we mean alloys between isomorphous elements or compounds.

It is not immediately obvious that the additional scattering of phonons in a solid solution will lead to an increase in the figure of merit. The mean free path of the charge carriers is usually greater than that of the phonons so one might expect the reduction of the mobility to be more noticeable than that of the lattice conductivity. It turns out, however, that the disorder produced on forming a solid solution may have little or no effect on the carrier mobility. It seems that the carrier scattering is not affected if the long range order is preserved. The phonons have a smaller wavelength and are more strongly affected by disturbances in the short range order.

Sometimes there is an effect on the mobility when a solid solution is formed. Airapetyants *et al* [10] suggested that the motion of electrons and holes can be associated with the sub-lattices of electropositive and electronegative atoms respectively. The carriers might then be scattered if there is a disturbance on the appropriate sub-lattice. There is some experimental evidence to support this view since, for example, it is common practice to use bismuth telluride–antimony telluride alloys as positive thermoelements and bismuth telluride–bismuth selenide alloys as negative materials. However, it cannot be claimed that the principle is generally valid. What is certainly true is the proposition that the ratio of mobility to lattice conductivity can be increased by employing solid solutions.

Plots of the reciprocal of the lattice conductivity against the concentration of the second component in certain solid solutions are shown in figure 5.4. The relative increase is greatest when silicon is added to germanium since both these elements have exceptionally high values of the lattice conductivity. Nevertheless, there is a useful increase in the lattice resistivity for alloys based on bismuth telluride and lead telluride despite the fact that both these compounds already have rather small lattice conductivities in the pure state.

In the next section we shall discuss the problem of point-defect scattering of phonons in more detail. Here we mention the simple empirical rule for determining the lattice conductivity of a solid solution $A_{1-x}B_x$ that was employed by Ioffe. According to this rule

$$\frac{1}{\lambda_L} = \frac{1}{\lambda_{L0}} + 4x(1-x)\left[\frac{1}{\lambda_{Lm}} - \frac{1}{\lambda_{L0}}\right], \tag{5.11}$$

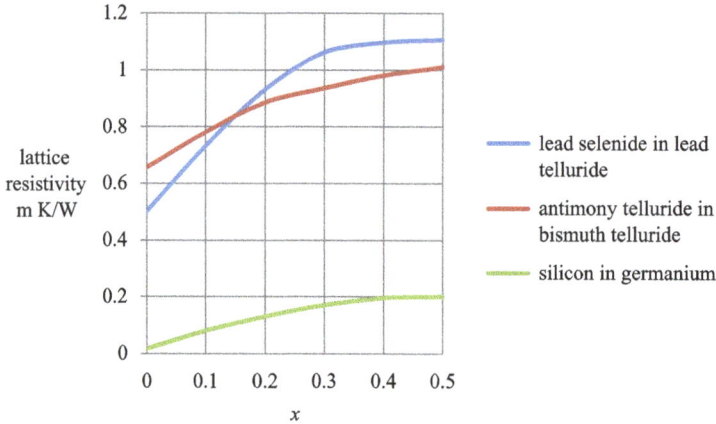

Figure 5.4. Lattice thermal resistivity for selected solid solutions at 300 K. x is the proportion of the added component.

where λ_{L0} is the lattice conductivity when $x = 0$ and λ_{Lm} is its value when $x = 1/2$. There are obvious difficulties when applying Ioffe's rule to solid solutions between components of very different lattice conductivities.

5.4 Mass-defect and strain scattering

We now consider the origin of the reduction of the lattice conductivity by point defects. Scattering will occur whenever there is a local change of the speed of sound. This may be due to a change of either the density or the elasticity. Thus, in a solid solution there will always be mass-defect scattering associated with the added atoms. There will also generally be strain scattering due to changes in the interatomic forces.

According to Rayleigh's theory, the scattering cross-section, σ, for the point defects is given by

$$\sigma = \frac{4\pi c^6 q_L^4}{9}\left(\frac{\Delta\chi}{\chi} + \frac{\Delta\rho}{\rho}\right)^2, \tag{5.12}$$

where c is the diameter of the defect and $\Delta\chi$ and $\Delta\rho$ are respectively the local changes of the compressibility and density. Rayleigh's theory is based on classical physics and is not expected to be valid for the phonons of high frequency. However, these phonons are very strongly scattered and, therefore, do not make much contribution to the thermal conductivity. Most of the heat transport is due to the low frequency phonons for which Rayleigh's theory is a good representation. We shall also use the Debye approximation for the vibrational spectrum for the same reason.

The problem that we have to face is that of including the normal processes in our calculations. This problem has been tackled rather successfully by Callaway [11]. He assumed that the action of the normal processes is to cause a disturbed phonon distribution to relax towards a distribution that still carries momentum.

We suppose that the equilibrium and disturbed distribution functions are N_0 and N respectively. Then the scattering processes will act on the distribution functions so that

$$\frac{\mathrm{d}N}{\mathrm{d}t} = -\frac{N_0 - N}{\tau_R} + \frac{N_N - N}{\tau_N}, \tag{5.13}$$

where τ_N is the relaxation time for those processes that conserve momentum and τ_R is that for those that do not. N_N is the distribution to which the normal processes on their own would relax. A key feature of Callaway's theory is the introduction of a constant vector l in the direction of the temperature gradient, which is such that $q \cdot l = h\upsilon$. The normal processes lead to a change in the distribution function given by

$$N_N - N_0 = \frac{q \cdot l}{kT} \frac{\exp(x)}{(\exp(x) - 1)^2}. \tag{5.14}$$

Since the vector l must be proportional to the temperature gradient we may write it as $-(\hbar/T)\beta v^2 \nabla T$, where β has the dimensions of time. We then find that there is an effective relaxation time τ_{eff}, which is given by

$$\frac{1}{\tau_{\mathrm{eff}}} = \frac{1/\tau_R + 1/\tau_R}{1 + \beta/\tau_N} = \frac{1/\tau_c}{1 + \beta/\tau_N}, \tag{5.15}$$

where τ_c would be the relaxation time if the normal processes were not momentum conserving.

The quantity β can be derived from the condition that the normal processes conserve momentum. It is found that this requirement is met if

$$\int_0^{\theta_D/T} \left(\frac{\beta}{\tau_N} - \frac{\tau_c}{\tau_N} - \frac{\beta\tau_c}{\tau_N^2} \right) \omega^4 \frac{\exp(x)}{(\exp(x) - 1)^2} \mathrm{d}x = 0, \tag{5.16}$$

where x is equal to $\hbar\omega/kT$.

The determination of β from equation (5.16) is usually not easy but there are some special cases for which the problem is simplified. Thus, if the scattering on point defects is very strong the relaxation time τ_R is very much less than τ_N and not much error is introduced if we treat the normal processes as if they were not momentum conserving. The high temperature approximation used by Parrott [12] is of particular interest in the context of thermoelectric materials and should hold when $T > \theta_D$.

Parrott assumed that the relaxation time should vary as ω^{-2} for umklapp and normal processes and as ω^{-4} for point-defect scattering. Also, $\hbar\omega/kT \ll 1$ for the whole vibrational spectrum at high temperatures. It is then found that

$$\frac{\lambda_L}{\lambda_{L0}} = \left(1 + \frac{5k_0}{9} \right)^{-1} \left[\frac{\tan^{-1}y}{y} + \left(1 - \frac{\tan^{-1}y}{y} \right)^2 \left(\frac{y^4(1 + k_0)}{5k_0} - \frac{y^2}{3} - \frac{\tan^{-1}y}{y} \right)^{-1} \right]. \tag{5.17}$$

In this equation k_0 is the relative strength of the normal to the umklapp processes and λ_{L0} is the lattice conductivity in the absence of point defects. The quantity y is defined from the relation

$$y^2 = \left(\frac{\omega_D}{\omega_0}\right)^2 \left(1 + \frac{5k_0}{9}\right)^{-1}, \tag{5.18}$$

and

$$\left(\frac{\omega_D}{\omega_0}\right)^2 = \frac{k}{2\pi^2 v \lambda_{L0} \omega_D A}, \tag{5.19}$$

where the relaxation time for point-defect scattering is equal to $1/A\omega^4$. k_0 can be found experimentally by measuring the thermal conductivity for one sample that contains imperfections and another that is free of defects.

The parameter A can be calculated when the scattering is due to density fluctuations on the basis of equation (5.12). It can be shown that

$$A = \frac{\pi}{2v^3 N} \sum_i \frac{x_i(M_i - M_{av})^2}{M_{av}^2}, \tag{5.20}$$

where x_i is the concentration of unit cells of mass M_i, M_{av} is the average cell mass, and N is the number of cells per unit volume.

The value of A for strain scattering is more difficult to predict. The local elastic constants change because of the misfit of the substituted atoms and because of the change of bond strength.

Experimental data on solid solutions show that the decrease of the lattice conductivity is often close to that predicted for mass-defect scattering. This is so for bismuth–antimony telluride and for germanium–silicon alloys. In other cases, for example in the PbTe–PbSe system, the observed lattice conductivity is smaller than expected for mass-defect scattering, suggesting that strain scattering is also significant.

5.5 Grain boundary scattering of phonons

There exists another way of reducing the lattice conductivity. It has long been known that λ_L becomes smaller as the size of a crystal is reduced [13] but boundary scattering was once regarded as a low-temperature phenomenon. However, it is now realised that this form of scattering can be a significant effect at high temperatures [14]. It is likely to have a stronger effect in solid solutions rather than pure elements or compounds even though the former have lower lattice conductivities.

If we use a classical model, the density of phonon modes increases as the square of the angular frequency, ω. On the other hand, the mean free path of the phonons in a pure crystal varies as ω^{-2}. Thus, we expect the contributions to the lattice conductivity to be about the same for all frequencies. It is the lowest frequency phonons that have the longest free path length and they can be affected by boundary scattering even when the average free path for all phonons is small. This is particularly so for solid solutions because mass-defect or strain scattering in these materials has the greatest effect at the largest frequencies. Incidentally, this justifies the use of a classical model, which is a very good approximation at low frequencies. The effects of the different forms of scattering are shown in figure 5.5. It is clear that the relative effect of boundary scattering is enhanced in solid solutions. Nevertheless, boundary scattering is large enough in thin single crystals of silicon for the thermal conductivity to be significantly reduced at ordinary temperatures.

The plot in figure 5.5 is obviously over-simplified since there will be phonon frequencies at which two types of scattering are contributing to the thermal resistance. These regions are rather narrow and the simple behaviour in figure 5.5 may not be too far from the true representation. If we accept this idea the expression for the lattice conductivity is

$$\frac{\lambda_L}{\lambda_S} = 1 - \frac{2}{3}\frac{\lambda_0}{\lambda_S}\sqrt{\frac{l_t}{3L}}. \tag{5.21}$$

In this equation λ_S is the lattice conductivity of a large crystal of the solid solution and λ_0 that in the absence of alloy scattering. The mean free path for phonon–phonon scattering is l_t and L is the effective grain size, which will presumably depend on the nature of the interfaces at the grain boundaries. l_t can be related to λ_0 using equation (5.2).

Equation (5.21) is applicable only when phonon–phonon scattering is strong enough for it to be dominant over part of the frequency range. If alloy scattering is very strong, equation (5.21) can be used until λ_L becomes equal to $2\lambda_S/3$. On the other hand, when alloy scattering is weak, equation (5.21) holds until λ_L is equal to $\lambda_S/3$. It is, therefore, useful when the boundary scattering is such that the lattice conductivity lies between one-third and two-thirds of its value for a large crystal.

It is interesting to discuss boundary scattering in material that has a large unit cell [15]. Until now the assumption has been made that the Debye model can be used

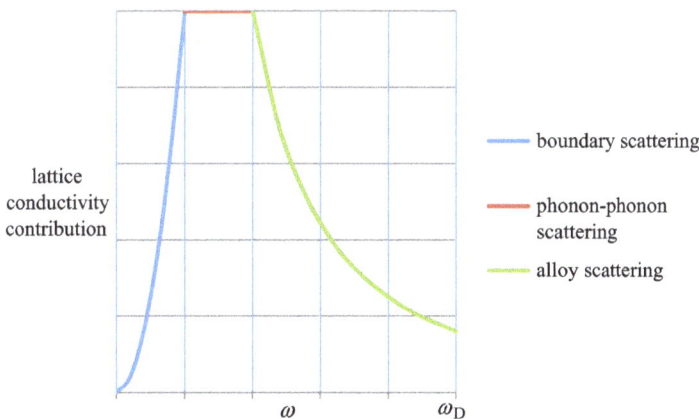

Figure 5.5. Schematic plot showing the contributions of different groups of phonons in a fine grained solid solution. The low frequency phonons are scattered by the grain boundaries and the high frequency phonons are subject to alloy scattering. Phonon–phonon scattering is dominant at intermediate frequencies.

and indeed this is satisfactory for the acoustic modes even when there are many atoms in the unit cell. However, we should also consider the effect of the optical modes.

It is still probable that most of the heat conduction is associated with the acoustic phonons since they have the largest group velocity. Nevertheless, we should not neglect the contribution of the optical phonons. It is easiest in this case to derive an expression for the change in the thermal conductivity due to boundary scattering rather than the relative effect as given in equation (5.21).

We introduce a frequency ω_0 equal to $(v/BL)^{1/2}$, at which umklapp scattering and boundary scattering are equally effective. We then determine the contribution, λ_a, to the lattice conductivity of the phonons up to this frequency with and without boundary scattering. If we make use of the Debye model for the specific heat,

$$\lambda_a = \frac{1}{9} vLc\omega_0^3, \tag{5.22}$$

with boundary scattering and

$$\lambda_a = \frac{1}{3} vLc\omega_0^3, \tag{5.23}$$

in the absence of boundary scattering. The difference between these two expressions yields the reduction in the thermal conductivity due to boundary scattering. This is

$$\Delta\lambda_L = \frac{2}{9} vLc_V\omega_0^3. \tag{5.24}$$

We can determine ω_0 if we make use of Lawson's relation, equation (5.7), to estimate the lattice conductivity when umklapp scattering is dominant. We find that

$$\omega_0 = \sqrt{\frac{a\omega_D^2 C}{9L a_T^2 \rho v^2 T}}, \tag{5.25}$$

where C is the total specific heat per unit volume.

This theory has been applied successfully [16] in explaining the observed thermal conductivity of the half-Heusler alloys with the formula $TiNiSn_{1-x}Sb_x$ for grain sizes between 1 and 10 μm.

5.6 Phonon drag

It has been assumed thus far that the electrons and phonons can be treated separately but this is sometimes invalid. When the flows of the two entities are linked we encounter what is known as the phonon drag effect [17]. It is a phenomenon that is usually observed at low temperatures, though it has been seen in semiconducting diamond above 300 K.

The Kelvin relations remain valid when phonon drag is the dominant mechanism and it is convenient to discuss the effect in terms of the Peltier coefficient. Suppose that there is a carrier concentration, n, and that an electric field, E, is applied. The carriers receive momentum at the rate $\mp ne\mathbf{E}$ per unit volume. This momentum may be lost in collisions with impurities or it may be passed on to the phonons, where it will remain until non-momentum conserving collisions occur. It will be assumed that a fraction x of the electronic collisions involve phonons. We shall set the phonon relaxation time for such collisions as τ_d. Then the excess momentum carried by the phonons is

$$\Delta p = \mp xne\tau_d \mathbf{E}. \tag{5.26}$$

The electric current density is $ne\mu\mathbf{E}$ and the heat flux per unit area is $v^2\Delta p$. Thus, the phonon drag Seebeck coefficient is

$$\alpha_d = \frac{\pi_d}{T} = \mp\frac{xv^2\tau_d}{\mu T}. \tag{5.27}$$

It must be emphasised that for the relevant phonons the relaxation time τ_d is strongly dependent on temperature. It has been suggested that τ_d is proportional to T^{-5} [17]. Thus, if we let the mobility vary as $T^{-3/2}$ we expect α_d to vary as $T^{-9/2}$. This accounts for the fact that phonon drag is essentially a low temperature effect. However, it will be noticed that equation (5.27) does not involve the carrier concentration and it might be thought that the power factor could be increased merely by introducing more charge carriers. Unfortunately, the phonon drag Seebeck coefficient becomes less than expected from equation (5.27) as n increases. This is partly due to scattering on the donor and acceptor impurities, which reduces x, but more importantly on a saturation effect. With an increased carrier concentration more and more of the momentum is fed back by the phonons into the electronic system. Herring showed that equation (5.27) should be modified to allow for the saturation effect so as to become

$$\alpha_d = \mp\left(\frac{\mu T}{xv^2\tau_d} + \frac{3nexv^2\tau_d}{N_d k\mu T}\right)^{-1}, \tag{5.28}$$

where N_d is the number of phonon modes interacting with the charge carriers. When the figure of merit is calculated using equation (5.28) for the Seebeck coefficient it appears that the phonon drag effect can only lead to values of zT that are significantly less than unity [18].

References

[1] Debye P 1914 *Vorträge über die kinetische theorie* (Teubner)
[2] Eucken A 1911 *Ann. Phys.* **34** 185
[3] Peierls R E 1929 *Ann. Phys.* **3** 1055
[4] Keyes R W 1959 *Phys. Rev.* **115** 564
[5] Dugdale J S and MacDonald D K C 1955 *Phys. Rev.* **98** 1751
[6] Lawson A W 1957 *J. Phys. Chem. Solids* **3** 154
[7] Lindemann F A 1910 *Phys. Z.* **11** 609
[8] Ioffe A V and Ioffe A F 1954 *Dokl. Akad. Nauk* **97** 821

[9] Ioffe A F, Airapetyants S V, Ioffe A V, Kolomoets N V and Stil'bans L S 1956 *Dokl. Akad. Nauk SSSR* **106** 981

[10] Airapetyants S V, Efimova B A, Stavitskaya T S, Stil'bans L S and Sysoeva L M 1957 *Zh. Tekh. Fiz* **27** 2167

[11] Callaway J 1959 *Phys. Rev.* **113** 1046

[12] Parrott J E 1963 *Proc. Phys. Soc.* **81** 726

[13] Casimir H B G 1938 *Physica* **5** 495

[14] Goldsmid H J and Penn A W 1968 *Phys. Lett.* **27** 523

[15] Sharp J W, Poon S J and Goldsmid H J 2001 *Phys. Status Solidi* A **187** 507

[16] Bhattacharya S, Skove M J, Russell M, Tritt T M, Xia Y, Ponnambalam V, Poon S J and Thadhani N 2008 *Phys. Rev.* B **77** 184203

[17] Herring C 1954 *Phys. Rev.* **96** 1163

[18] Keyes R W 1961 *Thermoelectricity: Science and Engineering* ed R R Heikes and R W Ure (New York: Interscience) p 389

Chapter 6

Materials for Peltier cooling

6.1 Bismuth telluride and its alloys

Our aim in this and the next chapter is to show how the principles that have been outlined previously are exemplified by existing thermoelectric materials. We shall first look at the most widely used of these, bismuth telluride, which was demonstrated to be suitable for thermoelectric refrigeration in the 1950s [1].

Crystals of bismuth telluride, with the formula Bi_2Te_3, have a hexagonal structure and display quite different mechanical properties in the a and c directions. Successive layers of bismuth and tellurium atoms are stacked as shown in figure 6.1. The bismuth and tellurium atoms are linked by strong ionic–covalent bonds but the neighbouring layers of tellurium atoms are held together by the weak van der Waals force. Consequently, crystals of bismuth telluride are easily cleaved in a direction perpendicular to the c axis.

The weak van der Waals bonding between the Te(1) layers is also evident from the diffusion of certain elements through bismuth telluride. Thus, the diffusion coefficient of copper in Bi_2Te_3 is several orders of magnitude greater in the plane of the a axes than it is in the c direction [2], as shown in figure 6.2. In fact, diffusion of copper along the cleavage planes is so rapid that the element can easily enter the lattice at room temperature. This can have a disastrous effect on the thermoelectric performance since copper is a donor impurity in Bi_2Te_3 and alters both the Seebeck coefficient and the electrical conductivity.

When bismuth is combined with tellurium in melt-grown material the maximum melting composition is not that with the stoichiometric proportions. At the composition with the highest melting point there is an excess of bismuth atoms so that undoped material is p-type rather than intrinsic. The excess of bismuth can be compensated by the addition of a donor impurity such as iodine. If sufficient iodine is added, the compound becomes n-type. Alternatively, the material can be made more strongly p-type by the addition of an acceptor impurity, for example, lead.

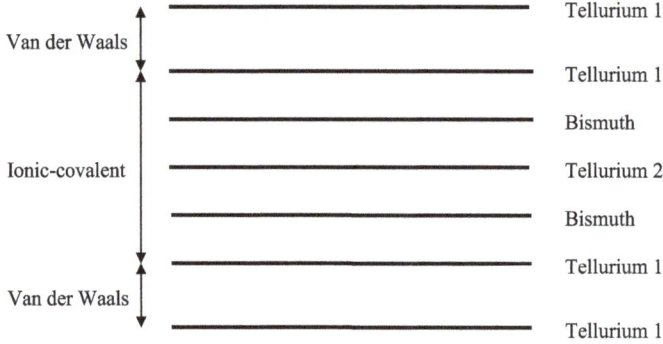

Figure 6.1. Layer structure of bismuth telluride.

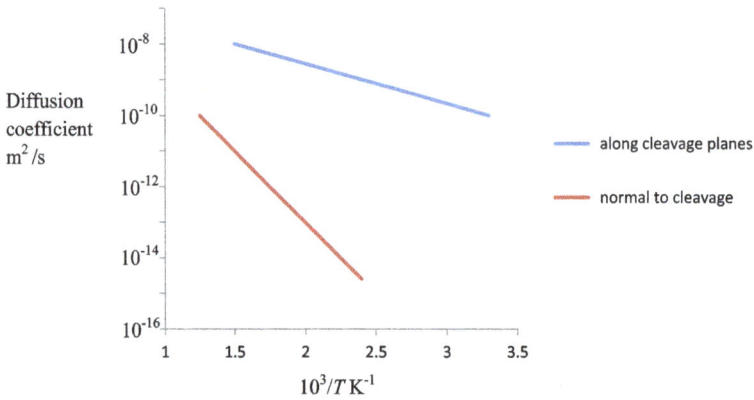

Figure 6.2. Diffusion coefficient for copper in bismuth telluride as a function of temperature.

The Seebeck coefficient of extrinsic bismuth telluride is isotropic but the electrical and thermal conductivities are both directionally dependent. The lattice conductivity is about twice as large parallel to rather than normal to the cleavage planes, with a similar anisotropy of the electrical conductivity in p-type samples. The electrical conductivity is about four times greater in the a direction than the c direction for n-type material. Thus, the figure of merit of randomly oriented polycrystalline n-type material is smaller than that of a properly oriented single crystal. The figure of merit is more or less the same for aligned and randomly oriented p-type material.

When bismuth telluride is grown from the melt, there is a tendency for the cleavage planes to lie parallel to the growth direction. Material produced in this way, even though not monocrystalline, yields thermoelements with the highest figure of merit whether they be p-type or n-type. In fact, aligned polycrystals are preferable to single crystals in that they have superior mechanical strength.

One of the favoured ways of producing thermoelectric materials is that of sintering of the powdered compound, usually by a hot-pressing technique [3]. This does not necessarily lead to grains that display any preferred orientation but alignment can be achieved by extrusion at an elevated temperature [4, 5]. Reduction of the grain size, of course, makes possible significant reduction of the lattice conductivity by boundary scattering of the phonons.

Most of the basic studies on the electronic properties of bismuth telluride have been performed using single crystals, carefully prepared to avoid cleavage. While aligned polycrystals are adequate when studying the thermoelectric parameters, they are of little use when determining the Hall and magnetoresistance coefficients. A detailed examination of the galvanomagnetic properties is one of the most straightforward ways of determining the band structure.

The Hall and the magnetoresistance effects in samples of bismuth telluride with different orientations were first investigated by Drabble and his colleagues [6, 7]. They found that the results could best be fitted by assuming that the energy minima for both the valence and conduction bands are located on the planes in wave vector space that contain the trigonal and bisectric directions. These directions are indicated in figure 6.3, which shows the first Brillouin zone for Bi_2Te_3. Symmetry then demands that there be six equivalent minima. The possibility of the minima lying on the faces of the Brillouin zone, which would give three minima, has been ruled out by further studies [8, 9].

The six-valley model for the band structure is probably one of the reasons that bismuth telluride has turned out to be such a good thermoelectric material. It allows the directionally dependent inertial effective mass to be small, leading to a high carrier mobility, while the density-of-states mass is large. Not so favourable is the rather small energy gap of about 0.13 eV, as revealed by the absorption edge in the infra-red region [10]. The details of the band structure are given in table 6.1. The similarity of the conduction and valence bands is consistent with the observation that the maximum figure of merit is not much different for n-type and p-type Bi_2Te_3. The electron and hole mobilities at 300 K are 0.12 and 0.051 m^2 (V s)$^{-1}$, respectively. The density of states effective masses for the two types of carrier are $0.58m$ and $1.07m$, giving the high values of 0.053 and 0.056 m^2 (V s)$^{-1}$ for $\mu(m^*/m)^{3/2}$. These values account for the fact that the power factors for both types of bismuth telluride are rarely equalled by other materials.

It has been found that the variation with temperature of the Seebeck coefficient is not quite consistent with a simple two band model. There is evidence [11] that there

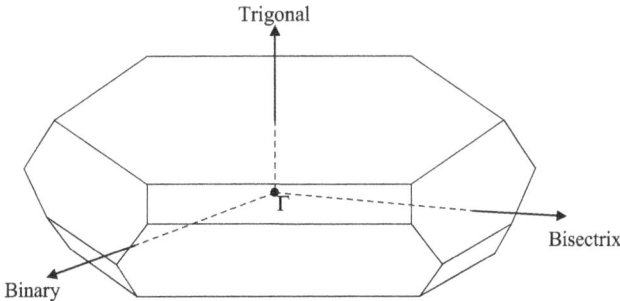

Figure 6.3. First Brillouin zone for bismuth telluride. A bisectrix direction lies perpendicular to the trigonal axis and one of the binary axes.

Table 6.1. Band structure of Bi_2Te_3. a_{ij}/m is a reciprocal effective mass tensor referring to axes lying in the reflection planes and an axis perpendicular to these planes.

Parameter	Valence band	Conduction band
Number of valleys	6	6
Location in k space	on reflection planes	on reflection planes
a_{11}	19.8	26.8
a_{22}	3.26	4.12
a_{33}	4.12	3.72
a_{23}	1.0	2.4
Energy gap (eV)	$0.13 - 9.5 \times 10^{-5}(T - 293)$	

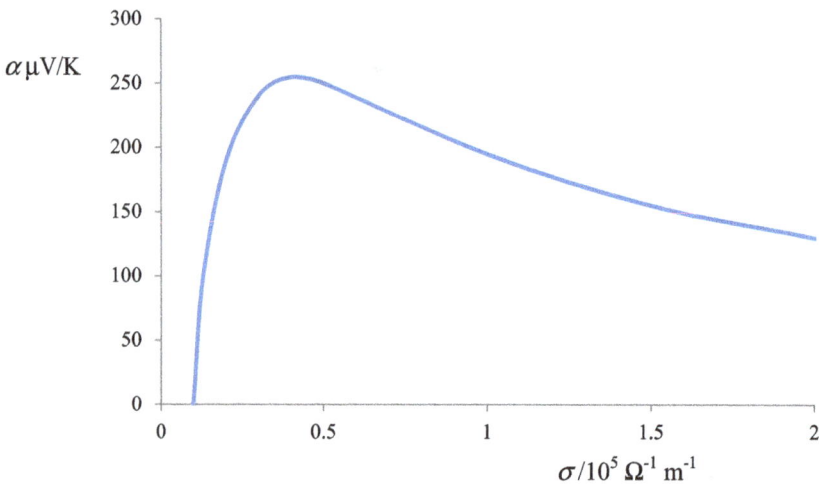

Figure 6.4. Seebeck coefficient at 300 K plotted against electrical conductivity for p-type bismuth telluride. The diagram extends into the region of mixed conduction.

is a second conduction band minimum with its edge about 0.03 eV above the main edge. Nevertheless, the band structure described by table 6.1 accounts reasonably well for the observed transport properties. The measurements on which it is based were carried out at liquid nitrogen temperature so as to enhance the galvanomagnetic coefficients over their room temperature values but it is thought that the band parameters at room temperature are not substantially different.

In figure 6.4 we show the variation of the Seebeck coefficient with electrical conductivity for p-type bismuth telluride at 300 K. The samples were oriented with current flow along the cleavage direction. The samples of higher conductivity were lead-doped and those of lower conductivity were doped with iodine. The behaviour of intrinsic samples, in which the iodine addition just balances the non-stoichiometry of undoped material, is indicated by the near-zero Seebeck coefficient on the left of the diagram. There will be a very similar plot, apart from the sign of the Seebeck coefficient, for n-type bismuth telluride with current flow perpendicular to the trigonal axis.

It will be seen from figure 6.4 that the Seebeck coefficient in the extrinsic region falls as the electrical conductivity rises. It turns out that the power factor reaches its maximum value when the Seebeck coefficient is close to 200 μV K^{-1}, as is clear from figure 6.5, in which the ratio of the power factor to its peak value is plotted against the Seebeck coefficient. The Seebeck coefficient at 300 K can never become higher than about 260 μV K^{-1} because of the narrowness of the energy gap.

The plot of thermal conductivity against electrical conductivity in figure 6.6 is particularly interesting. In the extrinsic range, that is for an electrical conductivity of the order of 10^5 Ω^{-1} m^{-1} or greater, the thermal conductivity rises with σ, the slope of the plot being close to what one expects from the Wiedemann–Franz law, bearing in mind that the degenerate approximation is inapplicable. However, for smaller electrical conductivities the thermal conductivity becomes much greater than expected for a single type of charge carrier [12]. One can obtain the lattice conductivity by extrapolation from the extrinsic region and one then finds that the electronic thermal conductivity behaves as shown in figure 6.7. In this diagram the ratio $\lambda_e/(k/e)^2\sigma T$ rises from its value of about 2 for a single type of carrier to about 16 in the intrinsic region. Thus, near-intrinsic material is unsuitable for thermoelectric energy conversion, not only because of its low Seebeck coefficient, but also because of its high thermal conductivity.

Although the simple compound, Bi$_2$Te$_3$ was used in the earliest successful demonstrations of thermoelectric refrigeration, the figure of merit can be increased through the use of solid solutions [13–15]. The improvement comes about mainly because of a decrease in the lattice conductivity. It is usual to use alloys of bismuth telluride with either antimony telluride or bismuth selenide. The variation of lattice conductivity with the bismuth telluride content is shown in figure 6.8.

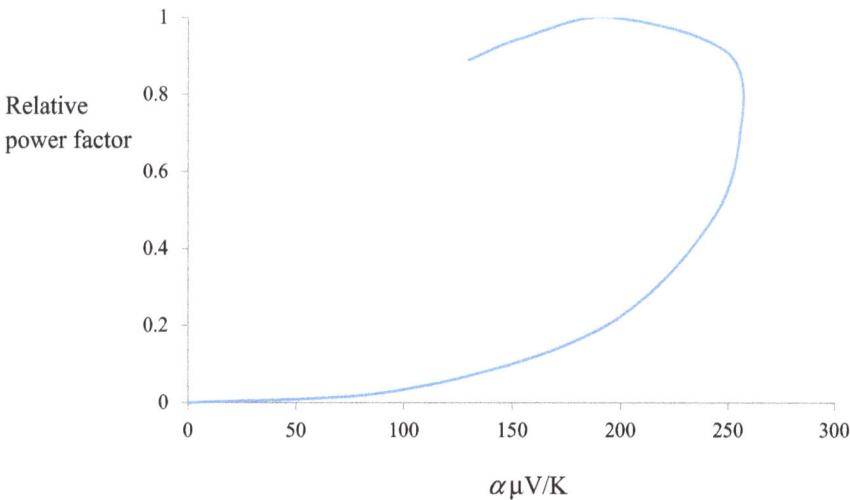

Figure 6.5. Ratio of power factor to its maximum value plotted against Seebeck coefficient for p-type bismuth telluride at 300 K.

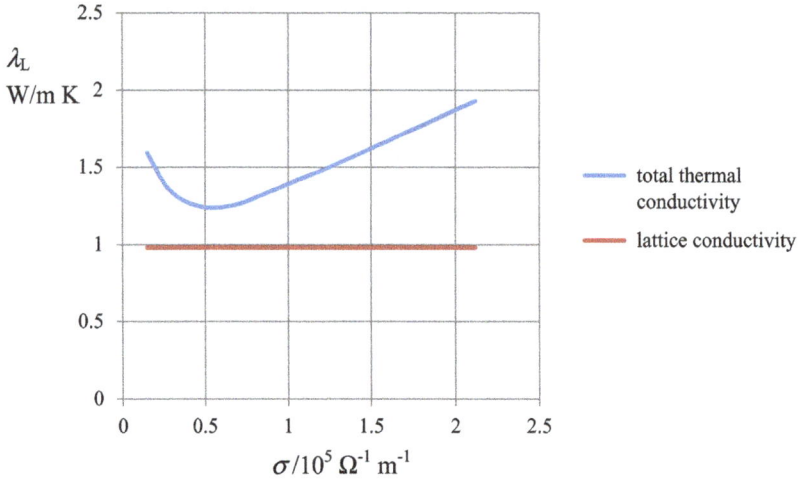

Figure 6.6. Plot of thermal conductivity against electrical conductivity for bismuth telluride at 300 K.

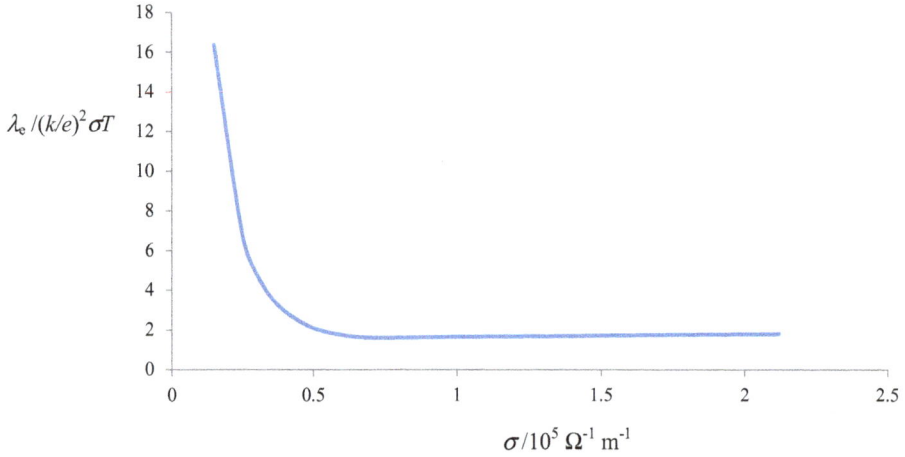

Figure 6.7. Plot of $\lambda_e/(k/e)^2\sigma T$ against electrical conductivity for bismuth telluride at 300 K.

The variation of lattice conductivity in the Bi_2Te_3–Sb_2Te_3 system is straightforward but there is less regular behaviour in the Bi_2Te_3–Bi_2Se_3 system. The unexpected maximum when the Bi_2Te_3 and Bi_2Se_3 concentrations are comparable may be an indication of some degree of ordering. It is found that the variation of energy gap with composition changes when the Bi_2Se_3 content reaches about one-third as shown in figure 6.9 [16]. It has been suggested that this may indicate that, as the selenide content rises, the selenium atoms first replace the Te(2) atoms and then the Te(1) atoms. As it happens, the electron mobility appears to fall significantly as the Bi_2Se_3 concentration rises above 20% and alloys with higher selenide concentrations than this are not yet employed in energy conversion. This may change when there is more interest in thermoelectric generation since the increased energy gap may assist in extending the operating temperature upwards.

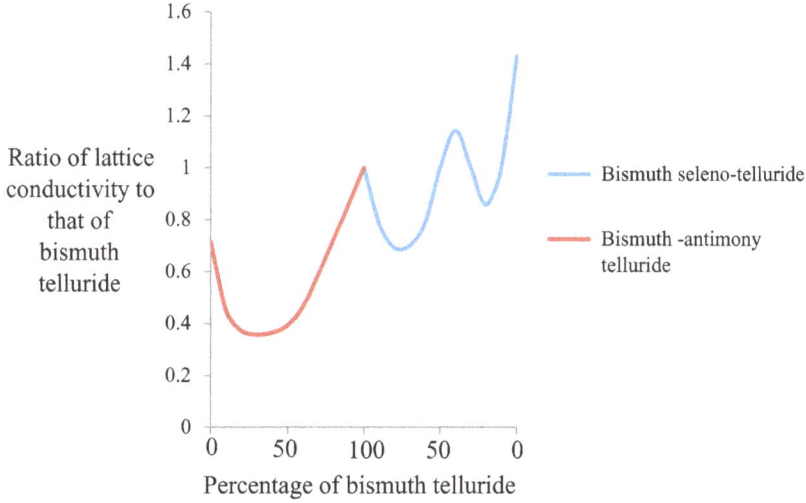

Figure 6.8. Lattice conductivity of bismuth telluride alloys at room temperature plotted against composition. The values are given as a proportion of the value for the compound Bi_2Te_3.

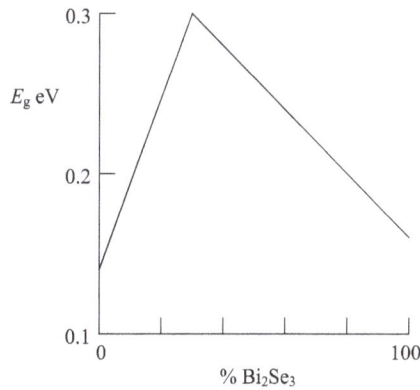

Figure 6.9. Plot of energy gap against composition in the Bi_2Te_3–Bi_2Se_3 system.

The typical compositions of the materials used in commercial thermoelectric modules are $Bi_{0.5}Sb_{1.5}Te_3$ and $Bi_2Te_{2.7}Se_{0.3}$ for the p-type and n-type branches respectively. For some time it was thought that the energy gap of Sb_2Te_3 must be too small to allow the concentration of this compound to exceed 75%. However, samples of Sb_2Te_3 with Seebeck coefficients of 240 $\mu V\ K^{-1}$ have now been reported so the energy gap in this compound should actually be comparable with that of Bi_2Te_3 [17].

Most of the applications of modules based on the bismuth telluride alloys have been in the field of refrigeration but there is growing interest in thermoelectric generation from heat sources at moderate temperatures. It is, therefore, important to know how the figure of merit will change as the temperature is raised [18, 19].

Let us, for the moment, ignore the effect of the minority carriers, though we know that they will become an increasing problem as the temperature is raised. If we assume that the carrier mobility varies as $T^{-3/2}$ and that the effective density of states is proportional to $T^{3/2}$, then both the optimum electrical conductivity and the optimum Seebeck coefficient should be more or less independent of temperature. For a pure crystal we would expect the lattice conductivity to be inversely proportional to the temperature but there will be a less rapid variation for the solid solutions. It is thought that the total thermal conductivity should vary little with temperature. This being so, zT should be approximately proportional to the temperature.

This favourable conclusion must be modified when one takes account of bipolar effects. Even for a wide gap one must increase the concentration of dopant to maintain the Seebeck coefficient at its optimum level. The dopant level must be increased still further when the energy gap is small. It would be helpful if the energy gap were higher than it is for Bi_2Te_3 and in this context the change on adding Bi_2Se_3 depicted in figure 6.9 gives cause for optimism. Likewise, the reports of improved Seebeck coefficients for Sb_2Te_3 suggest that one might be able to find an increased gap in the Bi_2Te_3–Sb_2Te_3 system. At present we can expect a small increase of zT for a modest rise of temperature above 300 K with a rather rapid decrease above, say, 400 K.

Experimental determinations of zT above room temperature show considerable differences between the observations of different workers. The highest values of zT are probably found for bulk nanostructures, that is material with nano-sized inclusions. Such materials have a value of about unity for zT at 300 K for n-type samples and a somewhat higher value for p-type specimens. In spite of the different methods of preparation all samples seem to display the same kind of behaviour. Typical results for zone-melted material are shown in figure 6.10. The compositions are not optimised for each temperature but it can be seen clearly that zT rises with temperature above 300 K but falls when the temperature is greater than about 450 K.

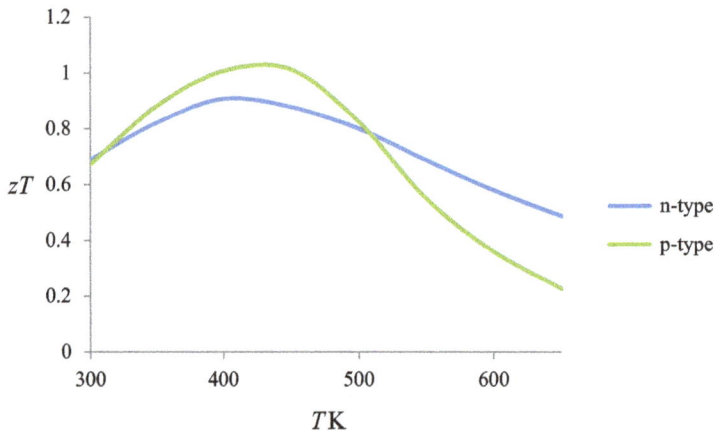

Figure 6.10. zT plotted against temperature for bismuth telluride alloys. The n-type material is $Bi_2Te_{2.4}Se_{0.6}$ and the p-type material is $Bi_{0.5}Sb_{1.5}Te_3$. Original data cited in [18].

Table 6.2. Melting temperatures of selected elements and compounds.

Material	Melting temperature (°C)
Bi	271
Zn_4Sb_3	563
Bi_2Te_3	573
Sb_2Te_3	629
Sb	631
Bi_2Se_3	710
SnTe	780
PbTe	917
Ge	937
PbSe	1065
PbS	1114
Si	1410

It is not certain what the upper bound to the temperature of operation may be but the melting temperature of 573 °C for bismuth telluride sets an obvious limit. Commercial modules made from bismuth telluride alloys can be operated continuously at 200 °C and intermittently at 230 °C. Table 6.2 shows the melting points of some of the materials of interest in the context of thermoelectric energy conversion.

6.2 Bismuth–antimony

It has long been recognised that Bi has reasonably good thermoelectric properties and these have been exploited in radiation thermopiles. In fact, if this element had an energy gap as large as, say, that of bismuth telluride it would be the best n-type material at 300 K. However, bismuth is actually a semi-metal and, even when heavily doped with a donor impurity, the thermoelectric properties are impaired by minority carriers.

Antimony is also a semi-metal but when it is alloyed with bismuth there is a range of composition for which there is an energy gap, albeit a small one. This allows Bi–Sb alloys to be used as low temperature n-type thermoelectric materials. Moreover, the high electron mobility in Bi and Bi–Sb means that there are large magnetic effects in quite modest fields. We shall discuss the longitudinal thermomagnetic effects in this chapter and deal with the transverse effects later.

The movement of the band edges when Sb is added to Bi is shown in figure 6.11 [20]. In Bi itself the light electron band overlaps a heavy hole band by about 50 meV. The light electron and hole bands are interchanged when more than 4% Sb is added and the overlap with the heavy hole band is removed at an Sb content of about 7%. Semiconducting properties remain until the Sb content reaches about 22% when a second heavy hole band overlaps the electron band. The maximum gap, which occurs when the Sb concentration is about 16%, is not large enough to prevent minority carrier problems at 300 K.

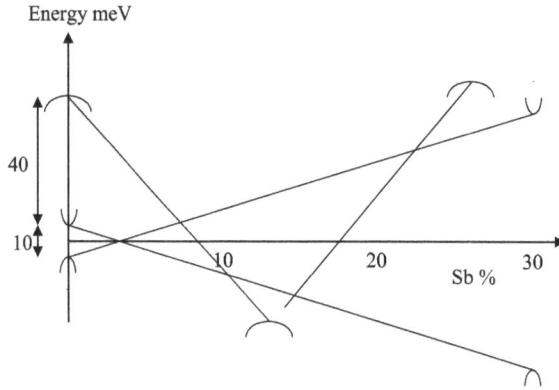

Figure 6.11. Schematic energy band diagram for Bi–Sb alloys (data from Lenoir *et al* [20]).

Bismuth has the same type of crystal structure as bismuth telluride and its lattice conductivity is smallest in the *c* direction. This also happens to be the direction of the highest electrical conductivity so there is no doubt that it is the preferred flow direction in a bismuth thermoelement. This requires the use of single crystals but they are brittle and are easily cleaved perpendicular to the *c* axis. Thus, there is some interest in polycrystalline Bi and Bi–Sb in spite of the fact that the figure of merit is much smaller than that in properly aligned single crystals.

The thermoelectric properties of single crystal bismuth in the *c* and *a* directions as a function of temperature are shown in figure 6.12 [21]. It will be seen that zT is an order of magnitude greater along the trigonal direction than in the plane of the binary axes. This is mainly because of the large difference between the Seebeck coefficients in the two directions.

Figure 6.12(c) also shows that the highest thermal conductivity is normal to the trigonal direction but it does not immediately allow us to separate the lattice and electronic components. One of the best ways of doing this is by applying a high transverse magnetic field. When the field, *B*, is such that $(\mu B)^2 \gg 1$, one might expect the electronic thermal conductivity to approach zero. However, this does not happen in the case of bismuth. The longitudinal temperature gradient produces a transverse electric field through the Nernst effect and this, in turn, leads to heat flow in the longitudinal direction through the Ettingshausen effect. This heat flow does not vanish, however large the magnetic field. Figure 6.13 shows how the total thermal conductivity in the binary direction changes with the strength of the magnetic field when this is directed along the trigonal and bisectrix directions [22]. That there is a difference between the high field values of λ is a clear indication that the electronic contribution to the thermal conductivity does not tend to zero. This difference does allow us to determine the lattice conductivity since the ratio of the magnetic field dependent part of the thermal conductivity in the two directions is easy to estimate from the known band parameters. As expected, the lattice conductivity is inversely proportional to the absolute temperature, with values of 2.9 and 2.0 W (m K)$^{-1}$ at 300 K in directions normal to and parallel to the trigonal axis respectively.

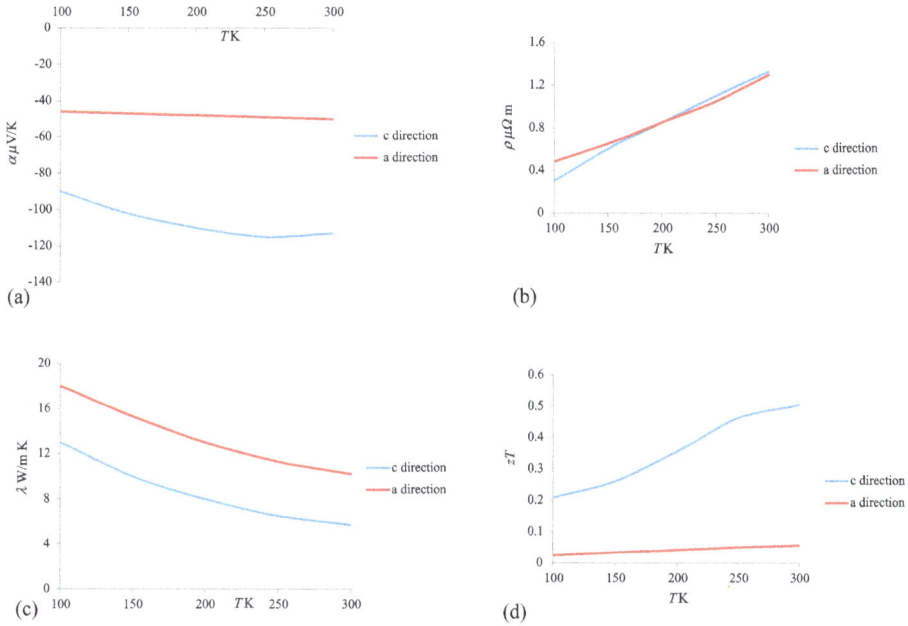

Figure 6.12. Thermoelectric properties of Bi parallel to and perpendicular to the trigonal axis: (a) Seebeck coefficient, (b) electrical resistivity, (c) thermal conductivity and (d) dimensionless figure of merit. (Based on the observations of Gallo *et al* [21].)

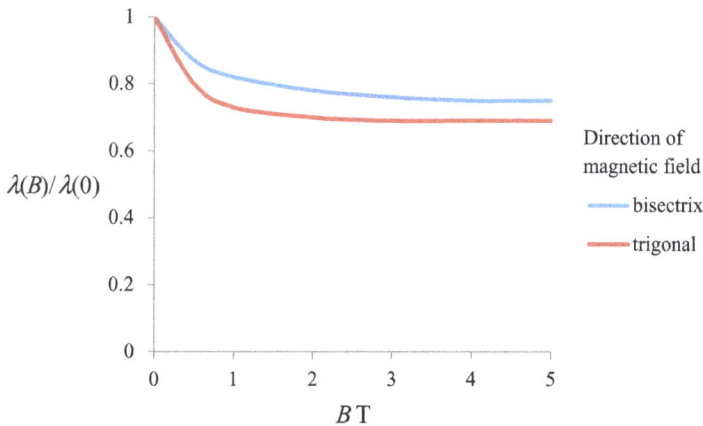

Figure 6.13. Thermal conductivity of Bi in the binary direction at 115 K. The ratio of the thermal conductivity to that in zero field is plotted against magnetic field strength. The magnetic field lies along the trigonal or bisectrix directions.

Bismuth displays another interesting phenomenon, known as the umkehr effect, in a transverse magnetic field. It is found that the Seebeck coefficient changes as the direction of the field is reversed. For example, it was observed by Smith and his colleagues [23] that one sample of Bi had a Seebeck coefficient of -150 μV K^{-1} with the magnetic field in a bisectrix direction. When the field was rotated through 180°, the Seebeck coefficient changed sign with a value of 170 μV K^{-1}. The umkehr effect can occur for any crystal with a suitable symmetry and with non-spherical constant-energy surfaces in wave vector space, but it is uncommon for it to be as profound as it is for bismuth.

It would appear that there should be an advantage, from the thermoelectric viewpoint, in using a Bi–Sb alloy rather than bismuth. The lattice conductivity for the alloy should be smaller and the appearance of an energy gap should reduce the adverse effect of the minority carriers. However, the addition of Sb seems to lower the value of $\mu(m^*/m)^{3/2}$. This is because the conduction band is non-parabolic close to its edge and this reduces the effective mass and there is no corresponding rise in the electron mobility [24]. Thus, the increase in zT on replacing Bi by Bi–Sb is rather small. The only advantage is the reduction in the lattice conductivity. Figure 6.14 shows λ plotted against magnetic fields for different Bi–Sb alloys at 80 K [25]. The temperature gradient is along a binary axis and the magnetic field in a bisectrix direction. It is evident that the lattice conductivity must be much less than the value of 11 W (m K)$^{-1}$ for Bi having the same orientation.

One can produce p-type Bi and Bi–Sb by doping with tin but the n-type material has a substantially higher figure of merit. The electron concentration can be increased by doping with tellurium but the highest n-type figure of merit has been observed for tin-doped material [26]. In fact, a single crystal of tin-doped $Bi_{0.95}Sb_{0.5}$ has yielded a value of z of about 3×10^{-3} K^{-1} in the trigonal direction over the range of temperature 120 to 280 K. This compares favourably with what can be achieved with n-type bismuth telluride but with that material aligned polycrystalline samples rather than single crystals are satisfactory. It is probable that Bi–Sb would be selected in preference to bismuth telluride only at low temperatures. It has been observed by Wolfe and Smith [27] that zT reaches the respectable value of about 0.4 for $Bi_{88}Sb_{12}$ at 80 K. Even though p-type Bi–Sb is inferior to n-type material it is still possibly good enough to be considered for use near liquid nitrogen temperature.

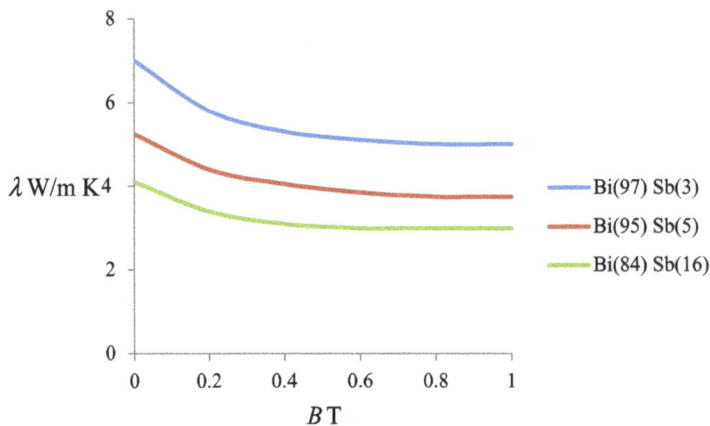

Figure 6.14. Thermal conductivity of Bi–Sb alloys at 80 K in a transverse magnetic field. The heat flow is parallel to a binary axis and the magnetic field is in a bisectrix direction. Observations of Cuff *et al* [25].

The longitudinal properties of both types of Bi and Bi–Sb can be improved by applying a transverse magnetic field and, at low temperatures, the field does not have to be impracticably large. Thus, at 160 K, Wolfe and Smith found that the figure of merit of $Bi_{88}Sb_{12}$ could be doubled in a field of 0.6 T.

References

[1] Goldsmid H J and Douglas R W 1954 *Br. J. Appl. Phys.* **5** 386
[2] Carlson R O 1960 *J. Phys. Chem. Solids* **13** 65
[3] Airapetyants S V and Efimova B A 1958 *Sov. Phys. Tech. Phys.* **3** 1632
[4] Seo J, Park K and Lee C 1998 *Materials Res. Bull.* **33** 553
[5] Pelletier R, Turenne S, Moreau A, Vasilevskiy D and Masut R A 2007 *Proceedings of the 26th International Conference on Thermoelectrics*, Jeju, Korea, IEEE 49
[6] Drabble J R and Wolfe R 1956 *Proc. Phys. Soc.* B **69** 1101
[7] Drabble J R, Groves R D and Wolfe R 1958 *Proc. Phys. Soc.* **71** 430
[8] Testardi L R, Stiles P J and Burshstein E H 1962 *Bull. Amer. Phys. Soc.* **7** 548
[9] Sehr R and Testardi L R 1963 *J. Appl. Phys.* **34** 2754
[10] Austin I G 1958 *Phys. Proc. Soc.* **72** 545
[11] Caywood L P and Miller G R 1970 *Phys. Rev.* B **2** 3209
[12] Goldsmid H J 1956 *Proc. Phys. Soc.* B **69** 203
[13] Rosi F D, Abeles B and Jensen R V 1959 *J. Phys. Chem. Solids* **10** 191
[14] Birkholz U 1958 *Z. Naturforsch.* A **13** 780
[15] Goldsmid H J 1961 *J. Appl. Phys.* **32** 2198
[16] Greenaway D L and Harbeke G 1965 *J. Phys. Chem. Solids* **26** 1585
[17] Hu L P, Zhu T J, Yue X Q, Liu X H, Wang Y G, Xua Z J and Zhao X B 2015 *Acta Mater* **85** 270
[18] Goldsmid H J 2014 *Materials* **7** 2577
[19] Goldsmid H J and Sharp J 2015 *Energies* **8** 6451
[20] Lenoir B, Dauscher A, Cassart M, Ravich Y I and Scherrer H 1998 *J. Phys. Chem. Solids* **59** 129
[21] Gallo C G, Chandrasekhar B S and Sutter P H 1963 *J. Appl. Phys.* **34** 144
[22] Uher C and Goldsmid H J 1974 *Phys. Stat. Solidi* A **65** 765
[23] Smith G E, Wolfe R and Haszko S E 1964 *Proceedings of the International Conference on Physics of Semiconductors, Paris* (Paris: Dunod) p 399
[24] Jain A L 1959 *Phys. Rev.* **114** 1518
[25] Cuff K F, Horst R B, Weaver J L, Hawkins S R, Kooi C F and Enslow G W 1963 *Appl. Phys. Lett.* **2** 145
[26] Jandl P and Birkholz U 1994 *J. Appl. Phys.* **76** 7351
[27] Wolfe R and Smith G E 1962 *Appl. Phys. Lett.* **1** 5

Chapter 7

Generator materials

7.1 IV–VI compounds and alloys

One particular compound that has a high mean atomic weight is lead telluride, PbTe. Its figure of merit is lower than that of bismuth telluride at 300 K but it has a larger energy gap, 0.32 eV, compared with 0.13 eV. Equally important is the fact that its melting temperature is more than 300° higher. Thus, although bismuth telluride is the best choice for thermoelectric generation with low source temperatures, lead telluride is superior above this range.

PbTe has cubic rock salt structure, so the thermoelectric properties are isotropic. It can be doped with donors, such as Zn, Cd, In, Bi or Cl, and acceptors that include Na, Au, Ti and O. The carrier concentration can also be adjusted by departures from stoichiometry [1]. The need to alter the concentration of dopant according to the operating temperature is clear from figure 7.1 [2], which shows zT plotted against temperature for two p-type samples of PbTe. The sample with a hole concentration of 2×10^{25} m^{-3} is superior to that with a concentration of 6.5×10^{25} m^{-3} at 300 K but the latter is the better above about 500 K.

When we compare the properties of PbTe and Bi$_2$Te$_3$ at 300 K, we find that PbTe has the high carrier mobilities of 0.16 m^2 (V s)$^{-1}$ and 0.075 m^2 (V s)$^{-1}$, for electrons and holes respectively. Although these mobilities are somewhat larger than the values for bismuth telluride, the density of states effective masses are only $0.21m$ and $0.14m$ so that $\mu(m*/m)^{3/2}$ is no more than 0.0154 m^2 (V s)$^{-1}$ and 0.0039 m^2 (V s)$^{-1}$ for n-type and p-type PbTe, respectively. These values are inferior to those for bismuth telluride. At least some of the difference may be attributed to the fact that both the valence and conduction bands in PbTe are of the 4-valley type whereas those in Bi$_2$Te$_3$ have 6 valleys.

The idea of using solid solutions to reduce the lattice conductivity was first demonstrated using alloys of PbTe with PbSe and there are a number of alloys of the IV–VI compounds that have been used in thermoelectric generation [2]. Usually the selected material comes from the general system Pb$_x$Sn$_{1-x}$Te$_y$Se$_{1-y}$ with the alloys

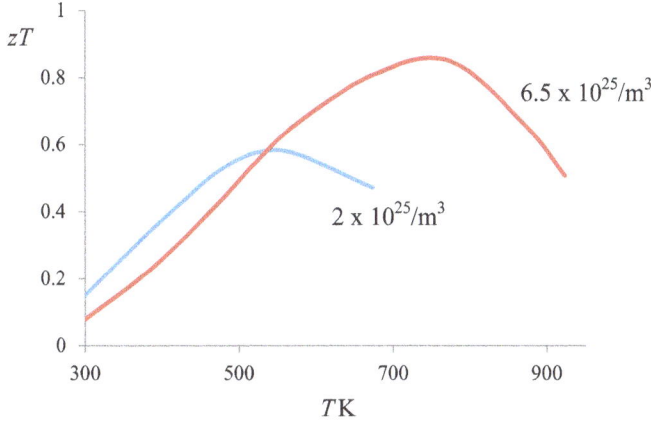

Figure 7.1. Plots of zT against temperature for two samples of p-type PbTe with different hole concentrations in the extrinsic range.

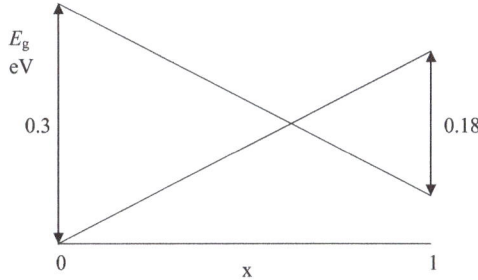

Figure 7.2. Schematic plot of energy gap against proportion of SnTe in $Pb_{1-x}Sn_xTe$. This diagram corresponds to a temperature of 12 K. The energy gap on the PbTe-rich side becomes larger as the temperature rises.

represented by $Pb_xSn_{1-x}Te$ often being preferred. Of course one must balance the reduction of the lattice conductivity with any change in the carrier mobility. There are also changes in the band gap that must be taken into account [3]. Thus, in $Pb_xSn_{1-x}Te$ the energy gap becomes smaller with increase of x until it falls to zero when x is equal to about 0.7. As shown in figure 7.2, the energy gap rises again as x becomes greater than 0.7 and there is an interchange between the valence and conduction bands. It is found that n-type $Pb_xSn_{1-x}Te$ yields zT equal to about unity at 500 K.

There is much interest in more complicated alloys based on the IV–VI compounds. The compositions containing Te, Ag, Ge and Sb are known by the acronym TAGS and are useful p-type generator materials [4]. They can be regarded as alloys of GeTe and $AgSbTe_2$. $AgSbTe_2$ has a rhombohedral structure and there is a phase transition when about 20% is added to GeTe. This means that the alloys between these two compounds are subject to considerable strain and, when the GeTe content lies between 80% and 85%, the lattice conductivity is exceptionally small. TAGS materials can be used in conjunction with n-type $Pb_xSn_{1-x}Te$.

PbTe has been used to show that not all doping agents are equivalent to one another. It seems that the addition of thallium gives rise to resonant levels [5]. If additional levels are located at an optimum distance above the band edge, the Seebeck coefficient for a given electrical conductivity is enhanced. As shown in figure 7.3, the variation of zT with temperature is quite different for PbTe doped with thallium and sodium.

Perhaps some of the most interesting observations on PbTe have been made using nanostructured material. Harman and his colleagues [6] have worked with PbTe/Te superlattices, measuring the Seebeck coefficient, the Hall coefficient and the electrical conductivity. The superlattice periods were between 15 and 30 nm. Figure 7.4 shows how the power factor was found to change with carrier concentration. The results for the superlattice are compared with those for bulk PbTe. There was a very significant increase in the power factor for the superlattice with carrier concentrations in excess of 10^{25} m^{-3}. This is one of the few cases in which it

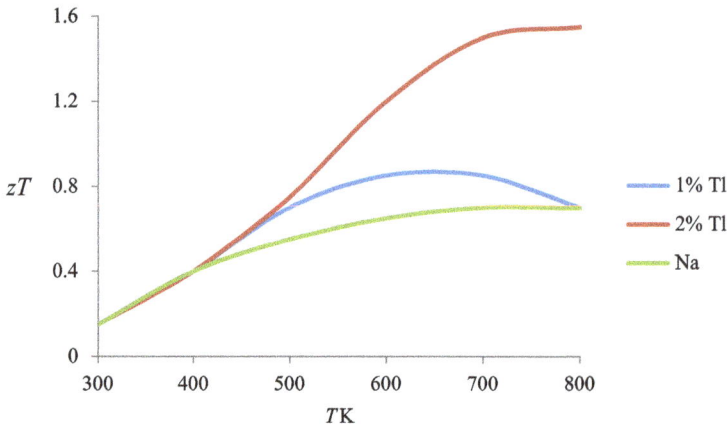

Figure 7.3. Schematic plots of zT against temperature for PbTe doped with Tl and Na. The former introduces resonant states above the band edge.

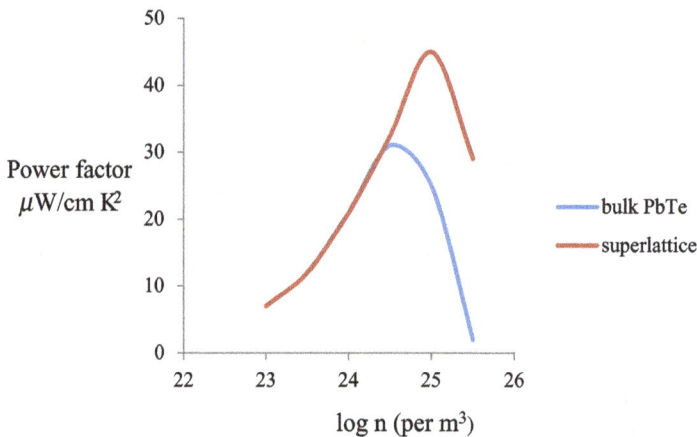

Figure 7.4. Variation of power factor with carrier concentration for a PbTe/Te superlattice.

7-3

has been found that the adoption of a nanostructure has improved the electronic properties as well as reducing the lattice conductivity.

Improvements have also been claimed for quantum dot superlattices made from PbTe and PbTe–PbSe [7]. Samples were produced by molecular beam epitaxy on BaF_2 substrates and were found to have a zT value of about 1.6 at 300 K rising to about 3 at 500 K. Presumably some credit for the high figure of merit must be attributed to a reduction in the lattice conductivity but it does seem that nano-structuring can also enhance the electronic properties in the PbTe system.

7.2 Silicon and germanium

Pure n-type silicon has a higher value of $\mu(m^*/m)^{3/2}$ at room temperature than bismuth telluride though the maximum power factor is not so good because the mobility is usually reduced at the optimum carrier concentration by impurity scattering. Impurity scattering is more evident for silicon because of its higher electron mobility. Also, the high dielectric constant in bismuth telluride shields the charge carriers from the Coulomb fields of the impurity ions. However, the main reason why silicon is not a good thermoelectric material is the very high lattice conductivity [8]. Its value of 145 W $(m\ K)^{-1}$ at 300 K is greater than the total thermal conductivity of many metals.

Germanium, too, has a large lattice conductivity, though at 64 m^2 $(V\ s)^{-1}$ it is somewhat smaller than that of silicon. Nevertheless, germanium–silicon alloys have been used as generator materials since they have a much lower lattice conductivity than either of the elements [9]. Figure 7.5 shows the variation of λ_L with composition at 300 K. Over much of the range of composition the lattice conductivity is of the order of 10 W $(m\ K)^{-1}$ but a rather lower value of about 5 W $(m\ K)^{-1}$ has been reported for $Ge_{0.3}Si_{0.7}$ [10]. In spite of the reduction in λ_L, the alloys of silicon and germanium have values of zT at ordinary temperatures that are too small for practical purposes, although it has been reported that a Si–Ge Peltier cooler has

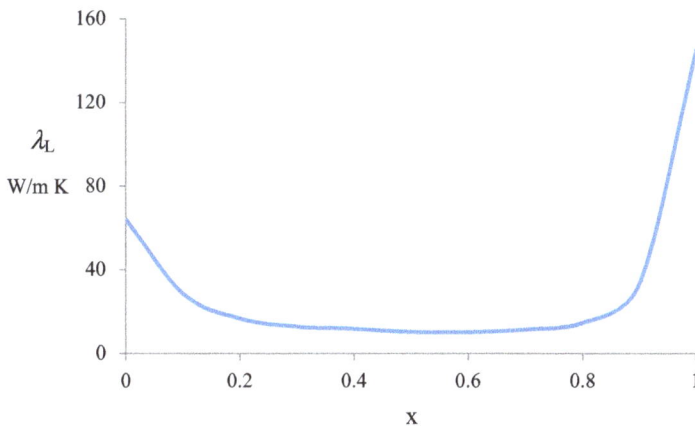

Figure 7.5. Plot of lattice conductivity against concentration of Si in Si_xGe_{1-x} alloys at 300 K. Schematic plot based on the data of Steele and Rosi.

been integrated into Si optoelectronic devices for dealing with hot spots [11]. The Si–Ge alloys come into their own as generator materials above 600 K with zT reaching about 0.5 and maintaining this value up to at least 1000 K [12]. The alloys seem to be quite stable at 1300 K. The energy gaps of 1.15 eV and 0.65 eV for Si and Ge, with intermediate values for the alloys, mean that minority carrier conduction has little effect in the heavily doped generator materials.

There is, of course, the possibility of further reducing the lattice conductivity by using fine grained material to introduce boundary scattering of the phonons. Experiments on pure silicon have shown that boundary scattering has an effect on the thermal conductivity at ordinary temperatures due to the substantial influence of the low frequency phonons [13]. There is an even greater effect in Si–Ge alloys because of the shift towards the low frequency region by alloy scattering. In spite of reservations expressed by Slack and Hussein [14], fine-grained sintered Si–Ge has proved to be a useful generator material.

The experience that has been gained on the use of silicon in microelectronic devices makes this element particularly suitable for the study of nanostructures. Despite the high lattice conductivity of bulk silicon, the value of λ_L for nanowires can be very small. For example, a silicon nanowire with a diameter of 52 nm was found to have a lattice conductivity of only 1.2 W (m K)$^{-1}$ [15]. Since the power factor for these nanowires was not much less than that of bismuth telluride, zT reached 0.6 at 300 K.

7.3 Phonon-glass electron-crystals

The figure of merit can be improved either by increasing $\mu(m^*/m)^{3/2}$ or by reducing the lattice conductivity. A low value of the lattice conductivity may be obtained in any particular material by the introduction of scattering centres for the phonons, for example impurities, dislocations or grain boundaries. Alternatively, one may select a material that has a small lattice conductivity even in its pure and perfect state. The lowest lattice conductivity that one might hope for is that of a glass or amorphous substance. Electrical conduction in a glass is unlikely to yield an acceptable power factor but Slack [16] proposed that materials might exist that would appear to be amorphous from the viewpoint of the phonons but crystalline in their electronic behaviour. These materials have come to be referred to as phonon-glass electron-crystals (PGECs).

PGECs may be found among crystals that have cage-like structures in which loosely bound impurity atoms may reside [17]. Two groups of crystals that embody the PGEC principles are the clathrates and the skutterudites.

The first clathrates to be studied were complexes of H_2O with trapped atoms or molecules and were known to have very low thermal conductivities. They have very large numbers of atoms in the unit cell, 46 H_2O molecules in Type I clathrates and 136 molecules in Type II clathrates. We are interested in those clathrates that exhibit semiconducting properties.

A characteristic of a PGEC is that the lattice conductivity should vary little with temperature except at low temperatures when the specific heat is temperature-dependent. It might typically have a value of about 0.5 W (m K)$^{-1}$, which is the value for amorphous Ge, at medium to high temperatures. One of the simplest clathrates is Cs_8Sn_{44} in which the Cs ions are rather large and do not fit loosely into the tin cages [18]. Consequently Cs_8Sn_{44} has the fairly high lattice conductivity of 10 W $(m\ K)^{-1}$ at about 7 K though the value becomes smaller as the temperature rises, becoming less than 1 W $(m\ K)^{-1}$ at 300 K. Other clathrates, such as $Sr_8Ga_{16}Si_{30}$, have smaller lattice conductivities which are far less dependent on temperature.

One of the more widely studied clathrates is $Ba_8Ga_{16}Sn_{30}$. A typical plot of Seebeck coefficient against temperature for this compound is shown in figure 7.6 [19]. The variation below 500 K is as expected for an extrinsic semiconductor, while the decrease above this temperature suggests the onset of minority carrier conduction. An increased donor concentration would take the Seebeck coefficient closer to the optimum value for high temperature operation. It has been reported that zT for copper-doped $Ba_8Ga_{16}Sn_{30}$ reaches 1.5 at 550 K, while, in p-type $Ba_8Ga_{15.9}Zn_{0.007}Sn_{30}$, zT is equal to 1.07 at 500 K [20]. There is no doubt, then, that the clathrates are useful thermoelectric generator materials.

The skutterudites are a class of material typified by $CoAs_3$ [21]. They are characterised by a unit cell that contains empty spaces. The unit cell consists of eight near-cubic arrangements of Co atoms. Six of these cubes contain almost square rings of As atoms leaving two voids that can be occupied by loosely bound atoms known as rattlers. It is these rattlers that are responsible for the low lattice conductivity. $CoSb_3$ has the relatively high temperature-dependent lattice conductivity of about 9 W $(m\ K)^{-1}$ at 300 K. In the widely studied skutterudite $La_{0.75}Fe_3CoSb_{12}$ the voids are partly filled with La atoms and the lattice conductivity is much lower. In the closely related partly filled skutterudite, $La_{0.75}Th_{0.2}Fe_3CoSb_{12}$, the lattice conductivity is only about 1 W $(m\ K)^{-1}$, a value that does not change much over a wide range of temperatures [22].

The electrons and holes in the skutterudites have a high effective mass and the Seebeck coefficient is large at carrier concentrations that would lead to metallic conductivity in most semiconductors. The figure of merit is small at room

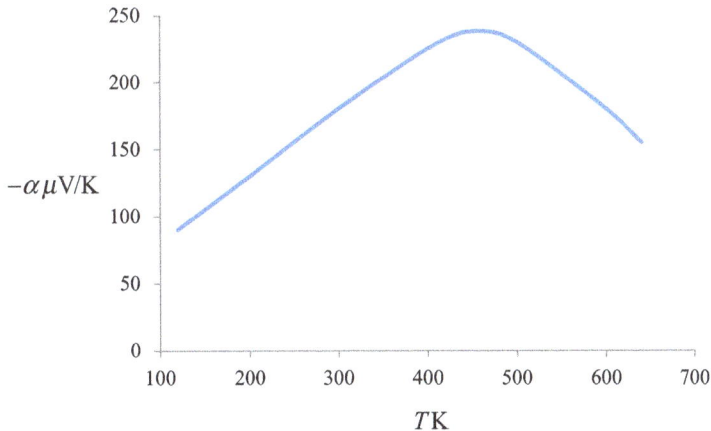

Figure 7.6. Seebeck coefficient of the clathrate $Ba_8Ga_{16}Sn_{30}$ plotted against temperature.

temperature but can become large as the temperature rises. Thus, zT for $Ce_{0.9}Fe_3CoSb_{12}$ reaches unity at a temperature of about 600 K. One of the features of the skutterudites is the large number of compositions that exist [23], many of them having similarly high values of zT.

7.4 Other thermoelectric materials

There are a number of other materials that display worthwhile thermoelectric properties. Thus, $FeSi_2$ has been used as a generator material even though it does not have a very large figure of merit. It is very stable, mechanically robust and composed of cheap and plentiful elements [24]. It serves to remind us that the figure of merit is not the only important parameter.

A material that does possess a large figure of merit is zinc antimonide [25]. The compound Zn_4Sb_3 exists in three crystalline forms. The β-phase is stable from 263 K to 765 K and is p-type with zT equal to 0.6 at about 500 K rising to about 1.3 at 700 K. There seem to be some problems with its mechanical stability because of the existence of the different phases. Another compound, ZnSb, is more stable and was used for the positive branches in some early generators. Its figure of merit was once thought to be rather low but recent work using different doping agents has yielded values of zT in excess of unity at 600 K.

Another group of materials that has attracted attention is the half-Heusler alloys. The Heusler alloy, Cu_2MnAl is ferromagnetic and has a structure in which the copper atoms form a cubic lattice with Mn and Al in alternate cells. In the half-Heuslers, half of the copper atoms would be missing. Among the half-Heusler alloys with the general formula MNiSn, M being Hf, Zr or Ti, several have useful n-type thermoelectric properties. In a typical compound, ZrNiSn, the power factor is satisfactory but the lattice conductivity is about 10 W (m K)$^{-1}$. This can be reduced by changing the composition to that of the alloy $Zr_{0.5}Hf_{0.5}NiSn$ [26]. Also, the high power factor is a consequence of a large density of states effective mass rather than a

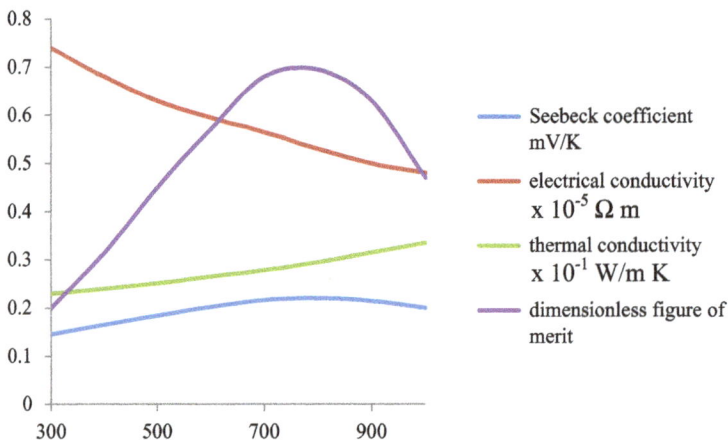

Figure 7.7. Thermoelectric properties of higher manganese silicide as a function of temperature.

high mobility so there is the likelihood of reducing the lattice conductivity by grain boundary scattering without affecting the electronic properties.

The thermoelectric properties of higher manganese silicide has been studied over a wide range of temperature and has been used as an example to illustrate the behaviour of a typical material. Figure 7.7 shows how the properties of a particular sample vary with temperature [27]. The maximum Seebeck coefficient of $210 \, \mu V \, K^{-1}$ at about 800 K is consistent with an energy gap of 0.32 eV. The fact that the Seebeck coefficient displays a maximum value indicates that an improved figure of merit at the highest temperature should be achieved by increasing the doping level. It is predicted that zT could rise to about 0.9 at 1000 K.

Finally, it may be mentioned that there are a number of organic materials that show some promise for thermoelectric applications. For one particular sample, designated as DMSO-mixed PEDOT-PSS, zT was reported to be 0.42 at room temperature [28]. The power factor has the rather modest value of $0.5 \, mW \, (m \, K^2)^{-1}$, which is only about one-tenth of the value for bismuth telluride, but this is balanced by the very small thermal conductivity of about $0.2 \, W \, (m \, K)^{-1}$. The ease of processing would make organic thermoelectrics very attractive.

References

[1] Allgaier R S and Scanlon W W 1958 *Phys. Rev.* **111** 1029

[2] Fano V 1994 *CRC Handbook of Thermoelectrics* ed D M Rowe (Boca Raton: CRC Press) p 257

[3] Melngailis I and Harman T C 1970 *Semiconductors and Semimetals* 5 ed A C Beer and R K Willardson (New York: Academic Press) p 111

[4] Skrabek E A and Trimmer D S 1994 *CRC Handbook of Thermoelectrics* ed D M Rowe (Boca Raton: CRC Press) p 267

[5] Heremans J P, Jovovic V, Toberer E S, Saramat A, Kurosaki K, Charoenphakdee A, Yamanaka S and Snyder G J 2008 *Science* **321** 554

[6] Harman T C, Spears D L and Walsh M P 1999 *J. Electron. Mats.* **28** L1

[7] Harman T C, Walsh M P, LaForge B E and Turner G W 2005 *J. Electron. Mats.* **34** L19

[8] Carruthers J A, Geballe T H, Rosenberg H M and Ziman J M 1957 *Proc. Roy. Soc.* A **238** 502

[9] Steele M C and Rosi F D 1958 *J. Appl. Phys.* **29** 1517

[10] Vining C B 1994 *CRC Handbook of Thermoelectrics* ed D M Rowe (Boca Raton: CRC Press) p 329

[11] Ezzahri Y, Zeng G, Fukutani K, Bian Z and Shakouri A 2008 *Microelectronics Journal* **39** 981

[12] Wood C 1988 *Rep. Prog. Phys.* **51** 459

[13] Savvides N and Goldsmid H J 1973 *J. Phys. C: Solid State Phys.* **6** 1701

[14] Slack G A and Hussain M A 1991 *J. Appl. Phys.* **70** 2694

[15] Hochbaum A I, Chen R, Delgado R D, Liang W, Garnett E C, Najarian M, Majumdar A and Yang P 2008 *Nature* **451** 163

[16] Slack G 1994 *CRC Handbook of Thermoelectrics* ed D M Rowe (Boca Raton: CRC Press) p 407

[17] Cahill D G, Watson S K and Pohl R O 1992 *Phys. Rev.* B **46** 6131

[18] Cohn J L, Nolas G S, Fessatidis V, Metcalf T H and Slack G A 1999 *Phys. Rev. Lett.* **82** 779

[19] Kuznetsov V L, Kuznetsova L A, Kaliazin A E and Rowe D M 2000 *J. Appl. Phys.* **87** 7871

[20] Saiga Y, Du B, Deng S K, Kajisa K and Takabatake T 2012 *J. Alloys and Compounds* **537** 303

[21] Nolas G S, Sharp J and Goldsmid H J 2001 *Thermoelectrics: Basic Principles and New Materials Developments* (Berlin: Springer) p 177

[22] Sales B C, Mandrus D, Chakoumakos B C, Keppens V and Thompson J R 1997 *Phys. Rev. B* **56** 15081

[23] Uher C 2001 *Recent Trends in Thermoelectric Materials Research I, Semiconductors and Semimetals* ed T M Tritt (San Diego: Academic Press) p 139

[24] Birkholz U, Gross E and Stöhrer U 1994 *CRC Handbook of Thermoelectrics* ed D M Rowe (Boca Raton: CRC Press) 287

[25] Zhua G, Liu W, Lan Y, Joshi G, Wang H, Chen G and Ren Z 2013 *Nano Energy* **2** 1172

[26] Uher C, Yang J, Hu S, Morelli D T and Meisner G P 1999 *Phys. Rev. B* **59** 8615

[27] Fedorov M I and Zaitsev V K 2006 *Thermoelectrics Handbook: Macro to Nano* ed D M Rowe (Boca Raton: CRC Taylor and Francis) p 31-1

[28] Kim G H, Shao L, Zhang K and Pipe K P 2013 *Nature Materials* **12** 719

IOP Concise Physics

The Physics of Thermoelectric Energy Conversion

H Julian Goldsmid

Chapter 8

Transverse flow and thermomagnetic effects

8.1 Advantages of the transverse thermoelectric effects

Until now we have discussed longitudinal thermoelectric devices. However, it is possible for a longitudinal flow of electric current to produce a transverse flow of heat and for a longitudinal temperature gradient to give rise to a transverse electric field. In other words, there can be transverse Peltier and Seebeck effects.

The utilisation of the transverse effects presents some problems arising particularly from the fact that the end contacts can act as transverse short circuits. Nevertheless, there are some advantages that stem from the heat and electrical flows being in perpendicular directions.

For example, suppose that we wish to make a thermal radiation detector. If we use the longitudinal Seebeck effect the output voltage will depend on the temperature difference, which, for a given heat flux, will depend on the length of the thermoelements. On the other hand, if we use a transverse device the output depends on the temperature gradient, which is independent of thickness. In other words, a transverse thermal detector can be very thin in the heat flow direction and this means that it can be very fast. The response time is proportional to the square of the length of the thermal path. A detector is based on the simple transverse thermoelectric generator shown in figure 8.1.

Let us suppose that the active area of the device has a cross-section $L_x L_z$ and a length in the heat flow direction, L_y. We find that the theory developed in chapter 2 for a longitudinal generator can be applied to the transverse device, though there are some significant changes that must be made. In the first place we are dealing with a single material rather than a thermocouple. Then, the expressions for the electrical resistance and the thermal conductance must be adapted to take account of the different flow directions for heat and electric charge. The electrical resistance in the x direction, R_x, is $\rho_x L_x / L_y L_z$ and the thermal conductance in the y direction, K_y, is $\lambda_y L_x L_z / L_y$. The transverse Seebeck coefficient, α_{xy} is defined as the ratio of the electric field E_x to the temperature gradient dT/dy, so the ratio of the thermoelectric

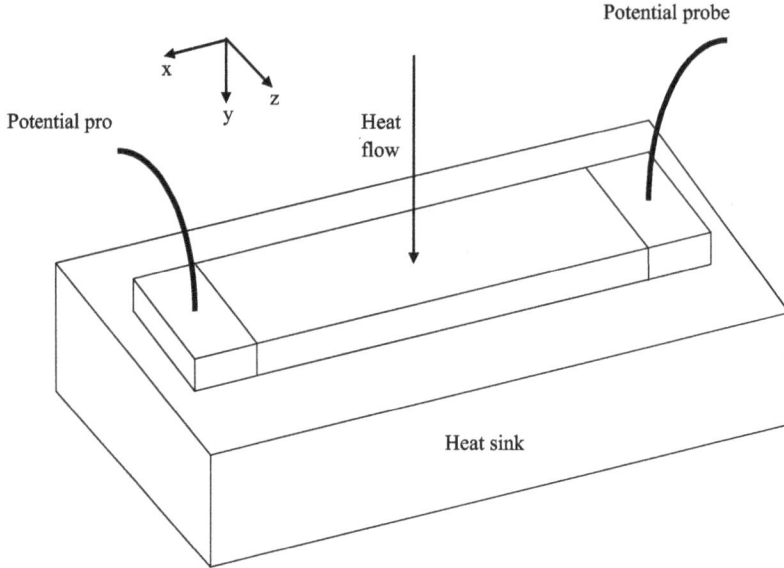

Figure 8.1. Transverse thermoelectric generator.

voltage in the x direction to the temperature difference in the y direction is $\alpha_{xy}L_x/L_y$. It is noted that the transverse figure of merit is $(\alpha_{xy}L_x/L_y)^2/R_xK_y$, which is equal to $\alpha_{xy}^2\sigma_x/\lambda_y$. In a longitudinal thermoelectric device the product of electrical resistance and thermal conductance in each element is equal to $\rho\lambda$ and is independent of the length and cross-section area but this is not true for a transverse generator or refrigerator. By using the transverse arrangement the electrical resistance and the thermal conductance can be adjusted more or less independently.

A single thermocouple is essentially a low-voltage high-current device that is matched to normal circuit requirements by using what was once called a thermopile, with many thermocouples connected in series and the heat flows in parallel. It is much easier to increase the voltage and decrease the current for a transverse device since one merely changes the length in the direction of the electrical flow.

The independence of the electrical resistance and thermal conductance greatly simplifies the design of a cascade. Let us consider a multi-stage refrigerator. We have already discussed the pyramidal arrangement when the longitudinal Peltier effect is employed. This allows the cooling power to be increased from stage to stage between the source and sink. With a transverse Peltier cooler the cooling power can be controlled by altering the thickness in the heat flow direction. Thus, a cascade may then take the form shown in figure 8.2(a). Moreover, by using the transverse Peltier effect one can realise the ideal of an infinite-staged cooler using an exponentially shaped element as shown in figure 8.2(b).

Let us consider a section of the device in figure 8.2(b) of thickness Δy at a distance y from the heat source. This section is one stage of the cascade. Its coefficient of performance is given by

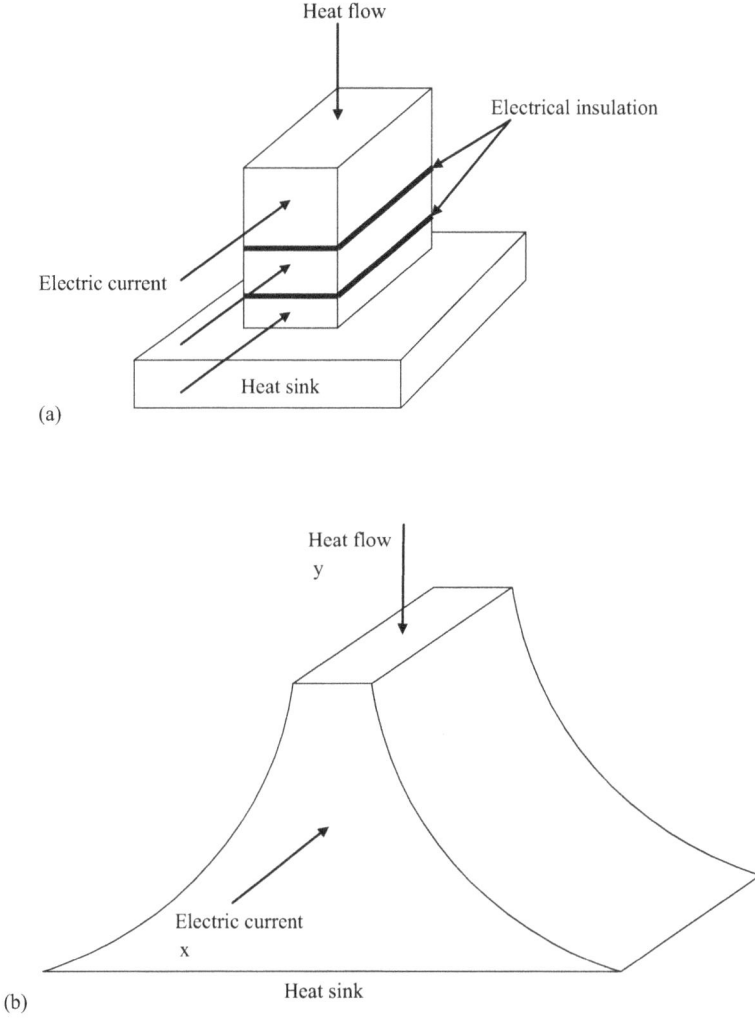

Figure 8.2. Transverse thermoelectric cascades. (a) A multi-staged refrigerator using elements of different thickness and (b) an exponentially shaped single element.

$$\phi_y = \frac{T}{\Delta T} \frac{(1 + z_{\text{trans}}T)^{1/2} - 1}{(1 + z_{\text{trans}}T)^{1/2} + 1}, \tag{8.1}$$

where z_{trans} is the transverse figure of merit and ΔT is the temperature difference across Δy. This means that the ratio of the heat flow emerging from the section to that entering it is

$$\frac{q_{y+\Delta y}}{q_y} = 1 + \frac{\Delta T}{T} \frac{(1 + z_{\text{trans}}T)^{1/2} + 1}{(1 + z_{\text{trans}}T)^{1/2} - 1}. \tag{8.2}$$

This, then, must be the ratio of the width in the z direction at $y + \Delta y$ to that at y. If we assume that $\phi_y T dT/dy$ is approximately constant we find that the length, L_z, in the z direction is

$$L_z = L_{z0} \exp\left(\frac{(1 + z_{trans}T)^{1/2} + 1}{T((1 + z_{trans}T)^{1/2} - 1)} y \frac{dT}{dy} \right). \tag{8.3}$$

This equation describes the shape of the infinite-stage transverse cooler. In practice, any tapering in the y-direction should give some improvement in performance compared with that of a device having a uniform width.

8.2 Synthetic transverse materials

There are few crystals that possess a significant directional dependence of the Seebeck coefficient. Even a substance like bismuth telluride, which displays a large anisotropy of the electrical and thermal conductivity, has the same Seebeck coefficient in the a and c directions unless both electrons and holes are present. The existence of the transverse thermoelectric effects was first observed by Korolyuk *et al* [1] for a single crystal of cadmium antimonide but such a material would have a very small transverse figure of merit. As we shall see later, substantial transverse effects in certain homogeneous crystals can be produced in a strong magnetic field, but otherwise we have to make use of two-phase materials if we wish to obtain worthwhile transverse figures of merit.

We consider the two-phase structure shown schematically in figure 8.3 [2]. The two phases are chosen so that they have different Seebeck coefficients and the layer thicknesses are such that $R_B \gg R_A$ and $K_A \gg K_B$, where R represents electrical resistance parallel to the layers and K is thermal conductance in the perpendicular direction. In the x_0 direction the thermal resistance will be dominated by material B whereas in the y_0 direction the electrical conduction will be dominated by material A. In the x_0 direction the Seebeck coefficient will be

$$\alpha_{x_0} = \frac{\alpha_A/K_A + \alpha_B/K_B}{1/K_A + 1/K_B}, \tag{8.4}$$

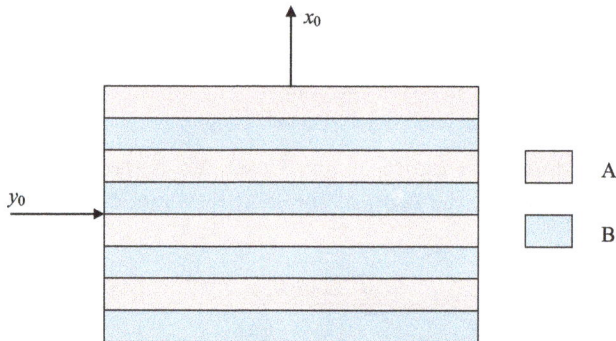

Figure 8.3. Synthetic transverse thermoelement made from layers of different conductors A and B.

and, when $K_A \gg K_B$, this approximates to α_B. In the y_0 direction the Seebeck coefficient is

$$\alpha_{y_0} = \frac{\alpha_A/R_A + \alpha_B/R_B}{1/R_A + 1/R_B}, \tag{8.5}$$

which is approximately equal to α_A.

Let us suppose that the thicknesses of the layers are d_A and d_B with d_B/d_A equal to n. Then equations (8.4) and (8.5) become

$$\alpha_{x_0} = \frac{\alpha_A/\lambda_A + n\alpha_B/\lambda_B}{1/\lambda_A + n/\lambda_B}, \tag{8.6}$$

and

$$\alpha_{y_0} = \frac{\alpha_A/\rho_A + n\alpha_B/\rho_B}{1/\rho_A + n/\rho_B}. \tag{8.7}$$

We can also obtain expressions for the effective thermal conductivities and electrical resistivities in the x_0 and y_0 directions. These are

$$\rho_{x_0} = \frac{\rho_A + n\rho_B}{n + 1}, \tag{8.8}$$

$$\rho_{y_0} = \frac{n + 1}{1/\rho_A + n/\rho_B}, \tag{8.9}$$

$$\lambda_{x_0} = \frac{n + 1}{1/\lambda_A + n/\lambda_B}, \tag{8.10}$$

and

$$\lambda_{y_0} = \frac{\lambda_A + n\lambda_B}{n + 1}(1 + Z_{AB}T_m). \tag{8.11}$$

It is noted that the thermal conductivity in the y_0 direction contains a term that includes a figure of merit Z_{AB}. This is because there will exist a Peltier effect due to circulating currents in the layers. Z_{AB} is the longitudinal figure of merit for a thermocouple between the materials A and B with branch dimensions related to n. It is given by

$$Z_{AB} = \frac{(\alpha_A - \alpha_B)^2}{(\lambda_A + n\lambda_B)(\rho_A + \rho_B/n)}. \tag{8.12}$$

In the synthetic two-phase material that has been described, the transverse thermoelectric coefficients would be zero with flow along the x_0 and y_0 directions. Non-zero values result from the inclination of a specimen at some angle ϕ to the y_0 direction. The transverse Seebeck coefficient is then

$$\alpha_{y_\phi x_\phi} = (\alpha_{x_0} - \alpha_{y_0})\sin \phi \cos \phi. \tag{8.13}$$

A condition for a large transverse Seebeck coefficient is that there should be a large difference between the longitudinal Seebeck coefficients of the two components.

The electrical resistivity and thermal conductivity of the composite material in the appropriate directions are

$$\rho_{x_\phi x_\phi} = \rho_{x_0} \cos^2 \phi + \rho_{y_0} \sin^2 \phi, \tag{8.14}$$

and

$$\lambda_{y_\phi y_\phi} = \lambda_{x_0} \sin^2 \phi + \lambda_{y_0} \cos^2 \phi. \tag{8.15}$$

We can combine these equations to obtain the transverse figure of merit

$$Z_{\text{trans}} = \frac{\alpha^2_{y_0 x_0}}{\lambda_{y_\phi y_\phi} \rho_{x_\phi x_\phi}}. \tag{8.16}$$

The optimum transverse figure of merit takes on a simple form when $\sigma_A \lambda_A \gg \sigma_B \lambda_B$. The expression is

$$Z_{\text{trans}}^{\max} = \frac{(\alpha_A - \alpha_B)^2}{\left(\{\lambda_A \rho_A\}^{1/2} + \left[\lambda_B \rho_B [1 + Z_{AB} T_m] \right]^{1/2} \right)^2}. \tag{8.17}$$

It is evident that this expression is almost the same as that for a longitudinal couple made from A and B apart from the term $(1 + Z_{AB} T_m)$ in the denominator. Thus, the transverse and longitudinal figures of merit may not be much different from one another. However, the transverse figure of merit will always be the smaller. The aim in selecting a pair of materials for a synthetic transverse thermoelement is to satisfy two conditions:

    (a) they should have a high figure of merit when used as a conventional thermocouple and

    (b) they should have widely different values for the product of electrical and thermal conductivity.

There is an optimum value for the angle ϕ which can be found from the equation

$$\tan \phi_{\text{opt}} = \frac{\sqrt{n}}{n+1} \left\{ \frac{\rho_A \lambda_B}{\rho_B \lambda_A} (1 + Z_{AB} T_m) \right\}^{1/4}. \tag{8.18}$$

Also, there is an optimum value for n, the ratio of the thicknesses of the layers. Thus,

$$n_{\text{opt}} \approx \left(\frac{2\lambda_B \rho_B / (\lambda_A \rho_A)}{1 + \lambda_B \rho_B / (\lambda_A \rho_A)} \right)^{1/2}, \tag{8.19}$$

but it does not seem that the value of n is critical. If the ratio of the thermal to electrical conductivity is similar for A and B the optimum value of n is close to unity and it is often satisfactory to make the layers of equal thickness.

There have been a number of reports of successful experiments on synthetic transverse thermoelements. Although it is difficult to make a satisfactory transverse device from p-type and n-type bismuth telluride alloys, it has been suggested that this might be possible if one were to combine crystalline bismuth seleno-telluride with porous bismuth–antimony telluride [3]. The porosity of the latter would allow the condition $\sigma_n \lambda_n \gg \sigma_p \lambda_p$ to be satisfied.

By accepting a somewhat lower longitudinal figure of merit one can also approach the condition $\sigma_A \lambda_A \gg \sigma_B \lambda_B$, using a semiconductor B combined with a metal A. This has been done by Kyarad and Lengfellner [4] who made a synthetic composite from n-type bismuth telluride and lead. They were able to obtain a temperature difference of 22° from the transverse Peltier cooling effect.

Gudkin *et al* [5] made a synthetic transverse thermoelement from bismuth–antimony telluride and the semi-metal Bi. When compared with Kyarad and Lengfellner's device, the longitudinal figure of merit for their pair of materials is greater but the condition $\sigma_A \lambda_A \gg \sigma_B \lambda_B$ is less well satisfied. Gudkin *et al* observed ΔT_{max} equal to 23° for a rectangular bar with a transverse figure of merit, z_{trans}, of 0.85×10^{-3} K^{-1}. They were able to increase ΔT_{max} to 35° by using a trapezoidal shaped element and could presumably have achieved a still greater value of ΔT_{max} by making a better approximation to the exponential shape of equation (8.3).

8.3 The thermomagnetic effects

There are four transverse effects that appear on the application of a transverse magnetic field. These effects are illustrated in figure 8.4. If the transverse electric fields and temperature gradients are in the directions shown, with the magnetic field directed away from the viewer, the four coefficients are regarded as positive.

The Hall coefficient, R_H, is defined as the ratio of the transverse electric field to the product of the longitudinal electric current density and the magnetic field B. The directions of the Hall field, the electric current and the magnetic field are mutually

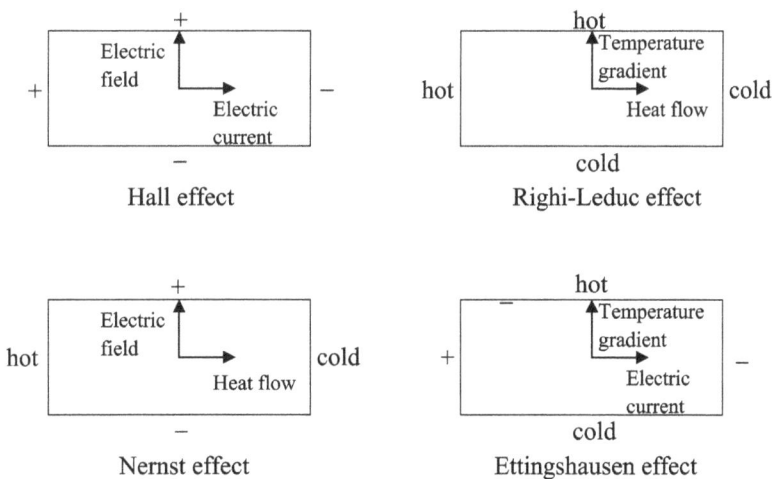

Figure 8.4. The transverse thermogalvanomagnetic effects. The magnetic field is directed into the page.

perpendicular. The Hall coefficient is a useful parameter in the study of semi-conductors since, when the charge carriers are all of one sign, it provides a measure of their concentration. The sign of the Hall coefficient determines the sign of the carriers.

The Nernst effect is the appearance of a transverse electric field on the application of a longitudinal temperature gradient. The Nernst coefficient, N, is defined by the relationship

$$N = \frac{dV/dy}{B_z dT/dx}.$$ (8.20)

The sign of the Nernst coefficient is independent of that of the carriers but, in an extrinsic conductor, it is a useful guide to the energy dependence of the relaxation time.

We are particularly interested in the Ettingshausen effect which, as shown in figure 8.4, is a transverse flow of heat resulting from a longitudinal electric current. The Ettingshausen coefficient is defined by

$$P = \frac{dT/dy}{i_x B_z},$$ (8.21)

where i_x is the electric current density in the x-direction.

The Ettingshausen and Nernst coefficients are related to one another in much the same way as the Peltier and Seebeck coefficients. However, there is a slight difference in the form of the relation because the Ettingshausen coefficient is defined in terms of a temperature gradient rather than a heat flow. Consequently the thermodynamic relationship for the transverse coefficients is

$$P\lambda = NT.$$ (8.22)

It will be noticed that this equation includes the thermal conductivity λ since it is this property that relates the temperature gradient to the heat flow.

Finally, the Righi–Leduc coefficient, S, is defined by

$$S = \frac{dT/dy}{B_z dT/dx}.$$ (8.23)

In principle, the Ettingshausen and Nernst effects can be used for refrigeration and generation, respectively [6]. However, the effects are rather small when there is only one type of carrier. The origin of the Ettingshausen effect is explained in figure 8.5 (a) where we arbitrarily suppose that the carriers are electrons. The magnetic field tends to move the electrons downwards but, because of the boundary conditions, there can be no net transfer of charge. If we assume that the low energy carriers are less strongly scattered than those of high energy it is still possible for the former to move downwards, with this motion balanced by the upwards movement of the high energy carriers. As one might imagine, this effect is not very large and it does not offer much promise for energy conversion.

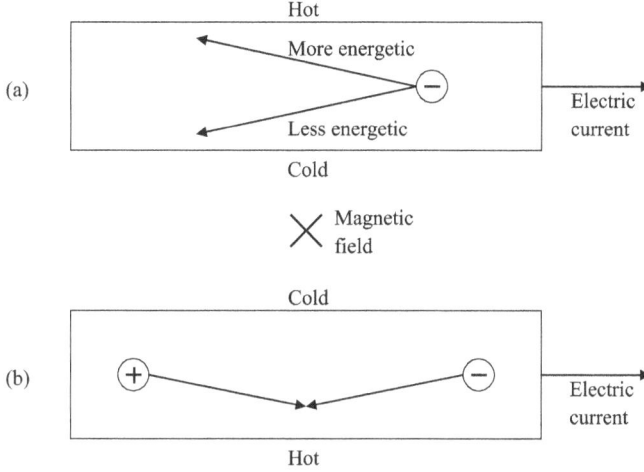

Figure 8.5. Origin of the Ettingshausen effect in (a) an extrinsic conductor and (b) an intrinsic conductor.

The Ettingshausen effect is much stronger when both electrons and holes are present. As shown in figure 8.5 (b), the presence of both types of carrier allows their transverse flow to take place without there being an overall electric current. It is this bipolar effect that is much more promising. We shall, therefore, derive an expression for the Nernst coefficient in a mixed conductor.

It is necessary that the longitudinal electron and hole currents balance one another so

$$i_{n,x} = -i_{p,x} = \frac{\sigma_n \sigma_p}{\sigma_n + \sigma_p}(\alpha_p - \alpha_n)\frac{\mathrm{d}T}{\mathrm{d}x}. \tag{8.24}$$

There will also be equal and opposite currents in the y direction due to the magnetic field, B_z. The electron current will be given by

$$i_{n,y} = \left(E_y + R_{H,n}i_{n,x}B_z\right)\sigma_n, \tag{8.25}$$

where E_y is the electric field due to the Nernst effect and $R_{H,n}$ is the partial Hall coefficient for the electrons. Likewise, the hole current will be

$$i_{p,y} = \left(E_y + R_{H,p}i_{p,x}B_z\right)\sigma_p. \tag{8.26}$$

By setting the transverse electric current equal to zero we find that

$$E_y = \frac{\left(R_{H,p}\sigma_p - R_{H,n}\sigma_n\right)\sigma_n\sigma_p}{(\sigma_p + \sigma_n)^2}(\alpha_p - \alpha_n)B_z\frac{\mathrm{d}T}{\mathrm{d}x}. \tag{8.27}$$

Thus, the Nernst coefficient is

$$N = \frac{\left(R_{H,p}\sigma_p - R_{H,n}\sigma_n\right)\sigma_n\sigma_p}{(\sigma_p + \sigma_n)^2}(\alpha_p - \alpha_n). \tag{8.28}$$

In a high magnetic field the Hall mobilities $R_{H,n}\sigma_n$ and $R_{H,p}\sigma_p$ become equal to the mobilities μ_n and μ_p as defined in chapter 4. Then equation (8.28) may be written as

$$N = \frac{(\mu_p - \mu_n)\sigma_n\sigma_p}{(\sigma_p + \sigma_n)^2}(\alpha_p - \alpha_n). \tag{8.29}$$

It may be noted that the product NB_z has the same dimensions as the Seebeck coefficient and has been referred to as the thermomagnetic power.

In order to determine the effectiveness of the transverse thermomagnetic effects for energy conversion we need expressions for the electrical and thermal conductivities in the applied magnetic field. The electrical conductivity is

$$\sigma = \frac{\sigma_n}{1 + \mu_n^2 B_z^2} + \frac{\sigma_p}{1 + \mu_p^2 B_z^2}. \tag{8.30}$$

A quantity of importance to us is the isothermal electrical resistivity, ρ_i, which is related to σ through the equation [7, 8]

$$\sigma = \frac{\rho_i}{\rho_i^2 + R_H^2 B_z^2}. \tag{8.31}$$

where R_H is the effective Hall coefficient which, at the limit of a very high magnetic field, becomes equal to $(1/R_{H,p} - 1/R_{H,n})^{-1}$.

It is difficult to derive an exact expression for the electronic thermal conductivity in a high magnetic field but a good approximation has been given by Tsidil'kovski [9]. His expression is

$$\lambda_e = \frac{\lambda_{e,n}}{1 + \mu_n^2 B_z^2} + \frac{\lambda_{e,p}}{1 + \mu_p^2 B_z^2} + \frac{\sigma_n\sigma_p(\alpha_p - \alpha_n)^2 T}{\sigma_n\left(1 + \mu_n^2 B_z^2\right) + \sigma_p\left(1 + \mu_p^2 B_z^2\right)}, \tag{8.32}$$

where $\lambda_{e,n}$ and $\lambda_{e,p}$ are the partial electronic thermal conductivities for the two types of carrier. This quantity tends to zero in a very high magnetic field under the condition that the transverse electric field is zero. When the transverse electric current is zero, the total thermal conductivity in an infinite magnetic field becomes equal to $\lambda_L(1 + Z_{NE}T)$ where Z_{NE} is the transverse thermomagnetic figure of merit.

Let us suppose that the concentrations of the electrons and holes are equal and that the magnetic field is large enough for $\mu_n^2 B_z^2 \gg 1 \ll \mu_p^2 B_z^2$. Then, from equation (8.29), the Nernst coefficient is

$$N = \frac{\mu_n\mu_p}{\mu_n + \mu_p}(\alpha_p - \alpha_n). \tag{8.33}$$

Also, in a very high magnetic field the Hall coefficient becomes equal to zero. Then the isothermal electrical resistivity is given by

$$\rho_i = \frac{1}{\sigma} = \left(\frac{\sigma_n(0)}{1 + \mu_n^2 B_z^2} + \frac{\sigma_p(0)}{1 + \mu_p^2 B_z^2}\right)^{-1}, \tag{8.34}$$

where $\sigma_n(0)$ and $\sigma_p(0)$ are the partial electrical conductivities in the absence of a magnetic field. We find that the thermomagnetic figure of merit of an intrinsic conductor in a very high magnetic field becomes

$$Z_{NE} = \frac{n_I e \mu_n \mu_p (\alpha_p - \alpha_n)^2}{(\mu_n + \mu_p)\lambda_L}, \tag{8.35}$$

where n_I is the intrinsic concentration of each type of carrier.

Of course, we cannot optimise the Fermi energy, as for a longitudinal thermo-electric device, since we have imposed the condition that the material is intrinsic. It is clear that the energy gap must be small or even slightly negative. If the energy gap were large enough for classical statistics to apply, the concentration of each type of carrier would be

$$n_I = 2\left(m_n^* m_p^*\right)^{3/2} \left(\frac{2\pi m k T}{h^2}\right)^{3/2} \exp\left(-\frac{E_g}{2kT}\right), \tag{8.36}$$

and, for a very large magnetic field,

$$(\alpha_p - \alpha_n) = \frac{E_g + 5kT}{eT}. \tag{8.37}$$

Thus, in the classical region

$$Z_{NE} \propto \left(E_g + 5kT\right)^2 \exp\left(-\frac{E_g}{kT}\right). \tag{8.38}$$

Then, assuming the classical approximation, we find that Z_{NE} falls continuously as the gap increases. It is expected that acoustic-mode lattice scattering of the carriers will predominate when the mobility is high, and for this condition Z_{NE} has its highest value when the energy gap is close to zero. If we suppose that the effective mass is the same for both types of carrier, the Nernst–Ettingshausen figure of merit reaches its maximum value when the valence and conduction bands overlap by about $2kT$ as shown in figure 8.6.

Ideally, the electrons and holes should have the same high mobility, μ. Then, equation (8.35) becomes

$$Z_{NE} = \frac{2 n_I e \mu \alpha^2}{\lambda_L}, \tag{8.39}$$

where α is the partial Seebeck coefficient for either carrier.

One of the major advantages of the transverse thermomagnetic effects over the longitudinal effects lies in the fact that a small or negative energy gap is no disadvantage. There is also an improvement by a factor of 2 that arises because the electrons and holes share a common lattice. Also, the partial Seebeck coefficients are usually increased by applying a high magnetic field. On the other hand, there are few materials that have a large enough carrier mobility for the condition $\mu^2 B^2 \gg 1$ to be

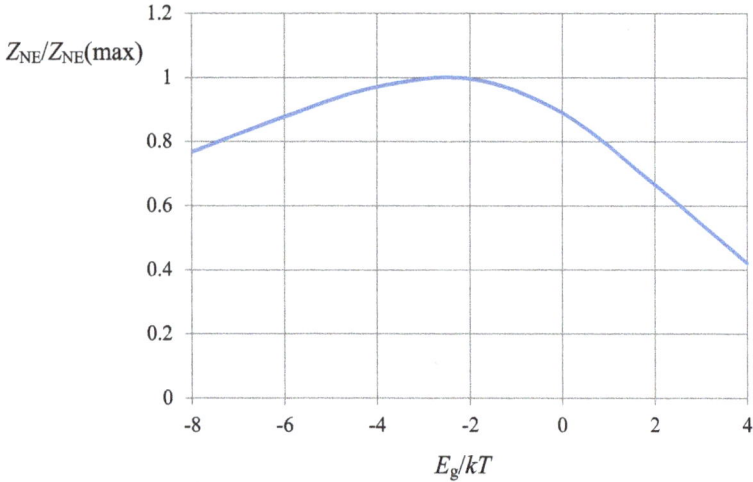

Figure 8.6. Dimensionless Nernst–Ettingshausen figure of merit as a fraction of its maximum value plotted against the energy gap. It is assumed that the electrons and holes have the same effective mass and that $\mu^2 B^2 \gg 1$ for both types of carrier.

approached. Typically, rare earth permanent magnets have a field strength of about 1 T so it seems that we need mobilities that are much greater than 1 m^2 (V s)$^{-1}$, if the device is to be practical. Indium antimonide has an electron mobility of about 7.7 m^2 (V s)$^{-1}$ at room temperature but the hole mobility is much smaller and the other parameters are not very favourable. The most promising results have been obtained using bismuth, particularly at low temperatures.

The basic parameters for bismuth at 80 K and 300 K are given in table 8.1 and the calculated values for the Nernst–Ettingshausen figure of merit in a high magnetic field at these temperatures are given in table 8.2. It will be seen that the value at 300 K for the best orientation is comparable with the longitudinal figure of merit of bismuth telluride alloys. However, at this temperature it is difficult to provide the necessary magnetic field. Thus, the thermomagnetic effects are most likely to be useful at low temperatures.

Yim and Amith [10] used the moderately high magnetic field of 0.75 T in their measurement of the transverse thermomagnetic figure of merit of bismuth between 70 K and 300 K. Their observation of the preferred orientation was consistent with table 8.2. However, because of the limited field strength, $Z_{NE}T$ at 300 K was no more than 0.025. At 80 K, $Z_{NE}T$ in the available field reached the value of 0.24, a value significantly larger than that in table 8.2.

One may be able to improve the high field value of $Z_{NE}T$ by using a Bi–Sb alloy rather than the element bismuth. As shown in figure 6.14, the lattice conductivity is substantially reduced when antimony is added to bismuth. Unfortunately, this beneficial effect is accompanied by a fall in the carrier mobility. This not only reduces $Z_{NE}T$ but it also makes the high field condition more difficult to achieve. Thus, it was found by Yim and Amith that a Bi$_{0.99}$Sb$_{0.01}$ alloy had a marginally higher value of Z_{NE} than Bi below about 130 K but, if the magnetic field was

Table 8.1. Parameters of Bi that are relevant to the transverse thermomagnetic figure of merit. The (1), (2) and (3) directions are parallel to the binary, bisectrix and trigonal.

Temperature	80 K	300 K
ne C m^{-3}	7.37×10^4	3.52×10^5
$\mu_n(1)$ m^2 (V s)$^{-1}$	55.7	3.18
$\mu_n(2)$ m^2 (V s)$^{-1}$	1.4	0.08
$\mu_n(3)$ m^2 (V s)$^{-1}$	33.3	1.9
$\mu_p(1) = \mu_p(2)$ m^2 (V s)$^{-1}$	12.4	0.77
$\mu_p(3)$ m^2 (V s)$^{-1}$	3.33	0.21
α_n μV K^{-1}	-100	-125
α_p μV K^{-1}	105	107
$\lambda_L(1,2)$ W (m K)$^{-1}$	11.0	2.9
$\lambda_L(3)$ W (m K)$^{-1}$	7.5	2.0

Table 8.2. Transverse thermomagnetic figures of merit, $Z_{NE}T$, in different orientations in Bi calculated for an infinite magnetic field.

Direction of temperature gradient	Direction of electric current	$Z_{NE}T$ at 80 K	$Z_{NE}T$ at 300 K
Bisectrix	Binary	0.0504	0.246
Binary	Bisectrix	0.0504	0.246
Binary	Trigonal	0.168	0.87
Bisectrix	Trigonal	0.0672	0.33
Trigonal	Binary	0.096	0.54
Trigonal	Bisectrix	0.096	0.54

restricted to 0.75 T, Bi had the higher value above this temperature. The highest value of $Z_{NE}T$ equal to about 0.5 was reached at a temperature of 150 K.

Perhaps the reduction in the mobility for the alloys can be avoided by careful preparation of the crystals so as to prevent imperfections resulting from impurities and constitutional supercooling. The growth of homogeneous single crystals of Bi–Sb is made difficult because of the wide separation of the liquidus and solidus in the phase diagram. Nevertheless, Horst and Williams [11] claimed to have obtained a value close to unity for $Z_{NE}T$ at 150 K for Bi$_{0.97}$Sb$_{0.03}$ in a field of 1 T.

References

[1] Korolyuk S L, Pilat I M, Samoilovich A G, Slipchenko V N, Snarski A A and Tsar'kov E F 1973 *Sov. Phys. Semiconductors* **7** 502
[2] Babin V P, Gudkin T S, Dashevskii Z M, Dudkin L D, Iordanishvili E K, Kaidanov V I, Kolomoets N V, Narva O M and Stil'bans L S 1974 *Sov. Phys. Semiconductors* **8** 478

[3] Goldsmid H J 2008 *J. Thermoelectricity* N1, 7

[4] Kyarad A and Lengfellner H 2006 *Appl. Phys. Lett.* **89** 192103

[5] Gudkin T S, Iordanishvili E K and Fiskind E E 1978 *Sov. Phys. Tech. Phys. Lett.* **4** 844

[6] O'Brien B J and Wallace C S 1958 *J. Appl. Phys.* **29** 1010

[7] Delves R T 1964 *Br. J. Appl. Phys.* **15** 105

[8] Horst R B 1963 *J. Appl. Phys.* **34** 3246

[9] Tsidil'kovskii I M 1962 *Thermomagnetic Effects in Semiconductors* (London: Infosearch) p 93

[10] Yim W M and Amith A 1972 *Solid State Electron.* **15** 1141

[11] Horst R B and Williams L R 1980 *Proceedings of the Third International Conference on Thermoelectrics, Arlington, Texas* (New York: IEEE) p 183

Chapter 9

Thermoelectric refrigerators and generators

9.1 Thermoelectric modules

In principle we can adapt a given thermocouple to provide a particular cooling power by adjusting the ratio of length to the cross-sectional area of the branches. However, in practice, this is not a reasonable approach unless the required cooling power is very small.

Consider equation (2.1) in which the rate of cooling of a single couple is expressed in terms of the electric current. The cooling power has its maximum value when

$$IR = (\alpha_p - \alpha_n)T_1. \tag{9.1}$$

Now $(\alpha_p - \alpha_n)$ will be of the order of 400 μV K^{-1} for an optimised thermocouple, while T_1 will be about 250 K. Thus, IR must be about 0.1 V. Ignoring the thermal conduction through the thermocouple, the cooling power per couple is then approximately 0.05I W, where I is expressed in amps. Then, if a cooling power of, say, 10 W is needed, the current will have to exceed 200 A. It is generally preferable to reduce the current to a few amps and to obtain the required cooling power by using several thermocouples in the form of a module. A thermoelectric module consists of a number of thermocouples connected in series electrically but with the heat flow in parallel.

A commercial module is shown in figure 9.1(a) and the components are shown schematically in figure 9.1(b). The thermoelements are linked to one another by copper connectors, which are in thermal contact with heat transfer plates that provide electrical insulation with a minimum of thermal resistance. The heat transfer plates may be metallised on the faces remote from the copper connectors. Aluminium oxide is an inexpensive material that is often used for these plates. Aluminium nitride and beryllia are better conductors of heat but the latter presents some health hazards. Diamond is equally good as an electrical insulator and has an even higher thermal conductivity and might be used in special applications.

(a)

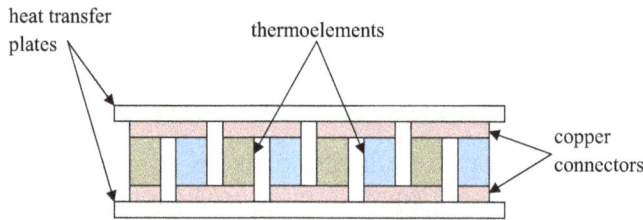

(b)

Figure 9.1. (a) Commercial thermoelectric module (courtesy of II-VI Marlow). (b) Schematic representation of section through a module.

The required cooling power and operating current will determine the ratio of length to cross-sectional area of each thermoelement. It might be thought that one should use the smallest possible volume of thermoelectric material but, as the length of each element is reduced, electrical and thermal transfer problems have to be faced.

One of these problems is the need to minimise the electrical resistance between the thermoelements and the copper links. This resistance can no longer be neglected when very short thermoelements are used. The design of very small modules has been studied by Semenyuk [1] who estimated the electrical contact resistance to be 0.84×10^{-10} Ω m^2. This suggests that the resistance at each contact would be about 1% of that of a 1 mm long thermoelement.

Parrott and Penn [2] have discussed the problem of contact resistance from the point of view of economy of material. As the length of the thermoelements is decreased, the relative effect of the contact resistance becomes greater. This means that ΔT_{max} becomes smaller. Parrott and Penn derived expressions for the ratio of cooling power to volume of thermoelectric material, with ΔT less than ΔT_{max}, under the conditions of either maximum cooling power or maximum coefficient of performance. There are generally two different thermoelement lengths from which to choose and one would usually opt for the greater of these as it would give the higher coefficient of performance.

Semenyuk was particularly interested in thermoelectric modules for cooling semiconductor laser diodes and other devices that involve a very high power density. He has shown that thermoelements of less than 1 mm in length can be used in specific applications. He found that there was little difference in performance between modules made from thermoelements of 200 and 150 μm in length. It was still possible to obtain a value of over 64° for ΔT_{max} with a thermoelement length of 130 μm. It may be noted that the heat transfer plates in the smallest modules were made from aluminium nitride rather than alumina so as to minimise the deterioration in performance due to poor heat transfer.

Let us now discuss the problem of heat transfer in parallel with the thermocouples using the simple model shown in figure 9.2 [3]. We suppose that the thermoelements are separated by some heat insulating material, which may well be air. As the separation between the elements is widened, so the area of the heat transfer plates becomes greater and this reduces the thermal resistance to the source and sink. However, the heat loss through the space between the elements is increased. We suppose that this space occupies g times the cross-sectional area of the thermoelements and that it has a thermal conductivity λ_I. The effective figure of merit is reduced to $Z/(1 + \lambda_I g/\lambda)$, where λ is the mean thermal conductivity of the thermoelectric materials. On the other hand, the thermal conductance of each of the end plates is increased from $K_c A$ to $K_c A(1 + g)$, where K_c is the thermal conductance per unit area of these plates.

The optimum value of g will depend on the specific requirements of the system. We suppose, for example, that the aim is to achieve the maximum temperature difference, ΔT^*_{max}, between the source and sink at zero load. ΔT^*_{max} will be less than the ideal value, ΔT_{max}, because of the thermal resistance of the end plates and the heat losses through the insulation between the thermoelements. When both these factors are taken into account we find that

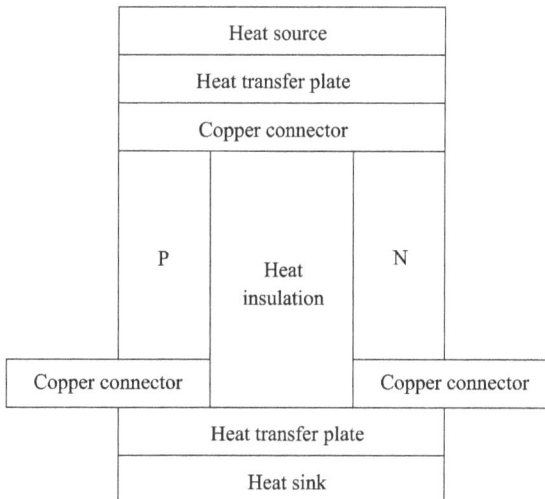

Figure 9.2. Simple model for the treatment of the heat transfer problem in a thermoelectric module.

$$\Delta T_{\text{max}}^* = \Delta T_{\text{max}} \left[\frac{1}{1+\lambda_I g/\lambda} - \frac{2\lambda T_2}{K_c L(1+g)T_1} \right]. \tag{9.2}$$

In figure 9.3 we show how $\Delta T_{\text{max}}^*/\Delta T_{\text{max}}$ varies with g in a typical case. The insulating medium has been assigned a thermal conductivity λ_I equal to 0.024 W (m K)$^{-1}$, which is the value for air. The heat transfer plates have been given a thermal conductance of 3×10^4 W (m^2 K)$^{-1}$, as expected for alumina of about 1 mm thickness. ZT has been assigned a value of unity. For this set of parameters it is apparent that the maximum temperature difference is achieved when the thermal insulation occupies about twice the space of the thermoelements. This maximum, however, is only about 92% of the value that it would have if there were no heat losses or end-plate thermal resistance. It is doubtful that there could be much improvement in the thermal insulation but the thermal resistance of the heat transfer plates could certainly be reduced.

9.2 Transient operation

It is usual to discuss the performance of thermoelectric coolers under steady-state conditions. It is also useful to know how rapidly a Peltier cooler will respond to changes in the load or the electrical power. Here we discuss the possibility of increasing the temperature depression by operating a thermocouple in a transient mode.

Babin and Iordanishvili [4] have determined the transient response of a couple consisting of two infinitely long legs. The junction is at $x = 0$ and the thermal load at this point is supposed to be negligible. Then the distribution of temperature along either leg must satisfy the equation

$$\frac{d^2 T}{dx^2} + \frac{i^2 \rho}{\lambda} = \frac{1}{\kappa} \frac{dT}{dT}, \tag{9.3}$$

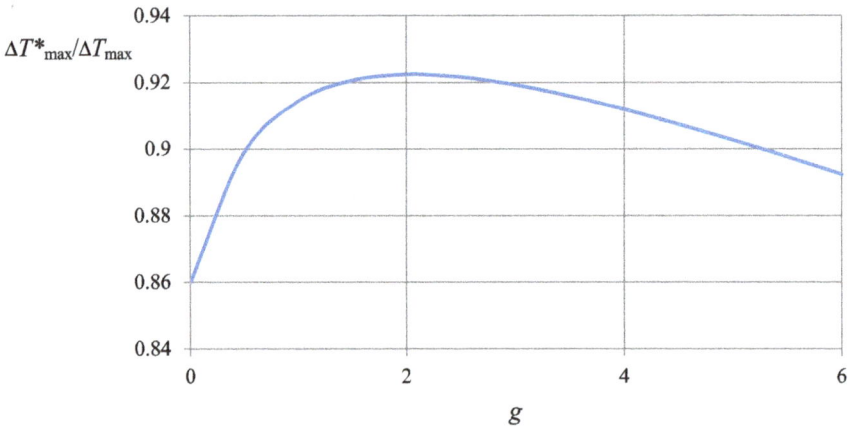

Figure 9.3. $\Delta T_{\text{max}}^*/\Delta T_{\text{max}}$ plotted against the filling factor g for typical values of the thermal conductivity of the insulating medium and the thermal conductance of the heat transfer plates.

where ρ, λ and κ are respectively the electrical resistivity, the thermal conductivity and the thermal diffusivity. We set the temperature as T_0 at all points when the time t is zero. Also dT/dx is equal to zero when $x = \infty$, while at $x = 0$, $\lambda dT/dx = \alpha IT$, where it is assumed that the Seebeck coefficient of each branch is $\pm\alpha$. Applying these conditions, the junction temperature T at time t is found from the equation

$$T_0 - T = T_0\left\{(1 - \exp(A^2)\mathrm{erfc}(A))\left(\frac{zT_0 + 1}{zT_0}\right) - \frac{2}{\pi^{1/2}}\frac{A}{zT_0}\right\}, \tag{9.4}$$

where $A = \alpha\kappa^{1/2}it^{1/2}/\lambda$. This equation shows that the temperature depression, ΔT, rises to a maximum value and then slowly decreases. If $zT_0 = 1$, the maximum depression is about $0.21T_0$ which may be compared with a value of $0.265T_0$ that would be achieved in the steady state with a heat sink. The electric current does not affect the maximum temperature depression but it does control the time scale.

The temperature depression can exceed the maximum steady state value if the current is increased after equilibrium is established. One may still use equation (9.3) but the temperature at $t = 0$, $x = 0$ is now given by ΔT_{max}. Equation (9.4) still holds if T_0 is replaced by $T_0 - \Delta T_{max}$. Babin and Iordanishvili showed that the temperature difference for a couple with z equal to 2.5×10^{-3} K^{-1} could be raised from 70° to 105° by this method.

Hoyos et al [5] imposed short current pulses on a bismuth telluride thermocouple through which a steady current was first applied. Their couples were tapered near the cold junction. The minimum temperature in the steady state was 220 K with the heat sink temperature equal to 290 K. A cold junction temperature of 175 K was achieved by applying pulsed currents equal to eight times the steady state value. For pulses of 50 ms duration, the recovery time was less than 2 s largely because of the tapering of the branches.

Woodbridge and Ertl [6] were able to enhance the temperature depression by using shaped current pulses. This technique was used by Landecker and Findlay [7] who increased the current continuously so that the Peltier effect could compensate the Joule heat as it reached the junction. Suppose that the current is proportional to $(t - t_0)^{-1/2}$ where t_0 is the time for which the pulse is applied and t is the time of observation. Then the temperature at time t is

$$T = \frac{ZT_0^2}{\pi}\ln\left(\frac{t - t_0}{t}\right). \tag{9.5}$$

Landecker and Findlay were able to observe temperatures as low as that of liquid nitrogen using a pulsed current of up to 100 A superimposed on a steady current of 5 A through a bismuth telluride couple.

It is noteworthy that pulsed cooling using either the transverse thermoelectric or thermomagnetic effects has the advantage that there is no thermal mass at a junction to slow down the response. Woodbridge and Ertl [8] used the Ettingshausen effect in bismuth to obtain 4° cooling below 80 K with a pulsed current as compared with only 1.2° for a steady current.

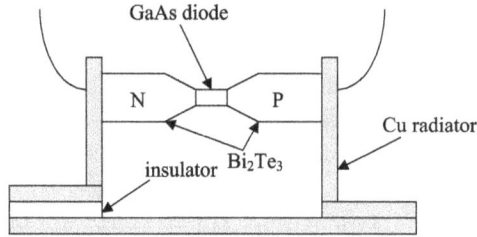

Figure 9.4. GaAs diode cooled by a pulsed bismuth telluride thermocouple. Schematic representation of Yamamoto's system [9].

Pulsed cooling has been put to practical use by Yamamoto [9] who passed the same current through a bismuth telluride thermocouple and a GaAs electroluminescent diode using the arrangement shown in figure 9.4. The doubling of the emitted radiation for pulsed operation indicated that the temperature had been lowered by an additional 50°.

9.3 Thermoelectric generators

The commercial exploitation of the thermoelectric effects has largely been devoted to refrigeration using the Peltier effect. However, the use of the Seebeck effect in the generation of electricity could eventually become much more important, particularly if materials with values of zT substantially greater than unity become available.

The efficiency of conversion of heat to electricity generally rises with the temperature difference between the source and the sink. When the temperature difference is large it is unfavourable to use a single material in each branch of a thermocouple. One way of solving this problem is by using multiple stages but a simpler approach lies in the use of segmented branches. One might expect to use two or more segments in each branch, with each segment having its optimised figure of merit. Sometimes this presents a compatibility problem.

For a thermoelectric generator made from single materials, the greatest efficiency is reached when the load is optimised. This means that the ratio, u, of electric current density to heat flux density has an optimum value, s, which is known as the compatibility factor. It is given by the relation

$$s = \frac{\sqrt{1 + zT} - 1}{\alpha T}. \tag{9.6}$$

It is not surprising that most thermoelectric materials have similar values of s since zT is usually close to unity and the optimum Seebeck coefficient is about $\pm 200\ \mu V\ K^{-1}$ but s is somewhat smaller for typical Si–Ge alloys. It is, therefore, difficult to match Si–Ge with the tellurides in segmented thermoelements.

This problem has been studied in some detail by Snyder and his colleagues [10, 11]. They have been able to explain the observation that the addition of a TAGS segment to a PbTe-based generator produced little extra power. Snyder has suggested that the compatibility factors for the materials in a segmented generator should not differ by a factor greater than two.

Efficiency is not usually thought to be an important factor when thermocouples are used for the measurement of temperature. However, it is of significance in radiation thermopiles, although it is not the only quantity of importance. The responsivity, R, is the ratio of the output voltage to the power of the incident radiation but, of greater relevance, is the specific detectivity, D^*. The detectivity is the reciprocal of the noise equivalent power, or the smallest detectable radiative input. The specific detectivity applies to unit surface area and band width and D^* allows us to compare different sensors.

Radiation thermopiles have been reviewed by Graf *et al* [12] who found the highest responsivity for a device made from bismuth–antimony telluride and a bismuth–antimony alloy. A larger specific detectivity equal to 88×10^5 m Hz$^{-1/2}$ was found for a Bi–Sb couple. The figure of merit is of some help in selecting thermocouples for radiation detection but geometric factors are of importance too. It is in this context that transverse devices have an advantage over normal thermocouples.

Turning to power generation, thermocouples possess the advantage of flexibility since they can operate over a wide range of temperatures if suitable materials are selected. Thermoelectric generators, drawing their heat input from radioactive sources, have been used in space vehicles for many years. Some of the early generators used thermocouples based on PbTe and its alloys but later generators working at higher temperatures made use of Si–Ge alloys, with efficiencies of up to 7%. In the context of space applications [13], efficiency is important since weight is significant and a higher efficiency means a smaller generator.

There are a number of heat sources that might be used with thermoelectric generators. Waste heat is available at many different temperatures and might need a range of segmented or staged thermocouples. Some possible sources have relatively low temperatures and can be exploited using thermoelectric generators made from the bismuth telluride alloys. The materials are little different from those used in Peltier cooling though the carrier concentrations have to be higher and the jointing techniques have to withstand higher temperatures. Thermoelectric modules suitable for operation in generators are available from the manufacturers of refrigeration devices. Low temperature sources include solar ponds and ocean thermal gradients. The low temperatures involved would make these sources difficult to utilise with other types of electrical generator.

9.4 Future prospects

The key to the advance of thermoelectricity lies in the development of materials with ZT values that are much greater than unity. In the field of refrigeration, the coefficient of performance at present is far short of what can be achieved with compression systems and, consequently, thermoelectricity is, for the most part, restricted to situations where the cooling power is low, say, of the order of less than 10 W. As ZT becomes greater, so Peltier cooling will come to be used in applications requiring larger cooling power. Our experience of thermoelectric systems has shown them to be unexcelled for reliability. For example, the behaviour of a thermoelectric air conditioning system installed on the French railways, with faultless operation over more than a decade, has shown what can be done [14]. Likewise, the improved efficiency that will result from higher values of ZT will make thermoelectric generators operating from low grade heat sources more economically viable.

From time to time predictions have been made of the maximum figure of merit that will ever be reached. Thus, many years ago [15] it was suggested that the best material in the future might combine the highest known value of $\mu(m^*/m)^{3/2}$, that is the value for electrons in bismuth, with a lattice conductivity of about 0.2 W $(m\ K)^{-1}$, that is typical of a glass. Then, if the energy gap were large enough to prevent any contribution from minority carriers, zT could become as large as 4.

The arguments that led to this conclusion are probably still valid today if we restrict ourselves to bulk materials. It is difficult to see any way of decreasing the lattice conductivity below a glass-like value but there remains the possibility of increasing $\mu(m^*/m)^{3/2}$, for example by adopting a nanostructure. In this context, it is interesting to note the theoretical calculations of Tan and his colleagues [16] on the thermoelectric properties of allotropes of carbon called graphyne and graphdiyne. Unlike the closely related allotrope, graphene, these layered structures are semiconductors with an energy gap of about 0.5 eV. The predicted value of zT is 5.3 at 580 K. Thus, there are now good reasons for hoping that thermoelectric materials with zT exceeding 4 will eventually be found.

References

[1] Semenyuk V 2006 *Proceedings of the Twenty Fifth International Conference on Thermoelectrics, Vienna* (New York: IEEE) p 322

[2] Parrott J E and Penn A W 1961 *Solid State Electron.* **3** 91

[3] Goldsmid H J 2016 *Introduction to Thermoelectricity* 2nd edn (Berlin Heidelberg: Springer) p 201

[4] Babin V P and Iordanishvili E K 1969 *Sov. Phys. Tech. Phys.* **14** 293

[5] Hoyos G E, Rao K R and Jerger D 1977 *Energy Convers.* **17** 45

[6] Woodbridge K and Ertl M E 1977 *Phys. Stat. Solidi* A **44** K123

[7] Landecker K and Findlay A W 1961 *Solid State Electron.* **3** 239

[8] Woodbridge K and Ertl M E 1978 *J. Phys. F: Metal Phys.* **9** 1941

[9] Yamamoto T 1968 *Proc. IEEE* **56** 230

[10] Snyder G J 2006 *Thermoelectrics Handbook: Macro to Nano* ed D M Rowe (Boca Raton: CRC Taylor and Francis) p 9–1

[11] Ursell T S and Snyder G J 2002 *Proceedings of the Twenty First International Conference on Thermoelectrics* (California: Long Beach) (IEEE: New York) p 412

[12] Graf A, Arndt M, Sauer M and Gerlach G 2007 *Measurement Sci. Technol.* **18** R59

[13] Abelson D D 2006 *Thermoelectrics Handbook: Macro to Nano* ed D M Rowe (Boca Raton: CRC Taylor and Francis) p 56-1

[14] Stockholm J G, Pujol-Soulet L and Sternat P 1982 *Proceedings of the Fourth International Conference on Thermoelectrics, Arlington, Texas,* (IEEE: New York) p 136

[15] Goldsmid H J 1964 *Thermoelectric Refrigeration* (New York: Plenum Press) p 130

[16] Tan X J, Shao H Z, Hu T Q, Liu G Q, Jiang J and Jiang H C 2015 *Phys. Chem. Chem. Phys.* **17** 22872